User Localization Strategies in the Face of Technological Breakdown

"Dr. Dorpenyo is to be applauded for highlighting an often forgotten issue of how technology issues in the Global South can inhibit social justice for users who rely on technology to participate in the democratic process. This is a timely and important material that will shape conversations on technology use in the fields of technical communication and rhetoric for a long time."
—Godwin Agboka, *Associate Professor of Technical Communication, University of Houston-Downtown, USA*

"*User Localization Strategies in the Face of Technological Breakdown* is a nuanced, insightful text that will be useful to technical communication researchers interested in theories and methodologies of localization, biometrics, and cross-cultural technical communication. A much-needed perspective from an important community that can completely transform the ways in which technical communicators think about technology design in both local and global contexts. This book makes powerful interventions in current conversations about decolonizing technical communication through social justice work."
—Laura Gonzales, *Assistant Professor of Rhetoric and Writing Studies at The University of Texas at El Paso, USA, and author of* Sites of Translation: What Multilinguals Can Teach Us About Digital Writing and Rhetoric

"Dorpenyo's work provides technical communicators with a deep and very privileged look into the fascinating world of technology transfer in Ghana. The story he tells of how biometrics were adapted by Ghana's election officials and voters is a case study for how to conduct analyses of 'user localization strategies' for our field."
—Tharon W. Howard, *Professor of Professional Communication and Rhetoric and Usability Testing Facility Director, Clemson University, USA*

"Dr. Dorpenyo's unique perspective and robust analysis of the adoption and use of biometric in Ghana's elections illustrates how users adapted this technology for their social, cultural, physical, and political contexts using linguistic, subversive, and user-heuristic localizations. This work, situated at the intersections of technical communication, civic engagement, social justice, user experience, and localization earns its significance by pointing out the importance of election technologies in non-western cultures and providing us with rhetorical localization strategies to consider within cultural technical communication."

—Michelle F. Eble, *Associate Professor of Technical and Professional Communication, East Carolina University, USA*

Isidore Kafui Dorpenyo

User Localization Strategies in the Face of Technological Breakdown

Biometric in Ghana's Elections

Isidore Kafui Dorpenyo
George Mason University
Fairfax, VA, USA

ISBN 978-3-030-26398-0   ISBN 978-3-030-26399-7 (eBook)
https://doi.org/10.1007/978-3-030-26399-7

© The Editor(s) (if applicable) and The Author(s), under exclusive license to Springer Nature Switzerland AG, part of Springer Nature 2020
This work is subject to copyright. All rights are solely and exclusively licensed by the Publisher, whether the whole or part of the material is concerned, specifically the rights of translation, reprinting, reuse of illustrations, recitation, broadcasting, reproduction on microfilms or in any other physical way, and transmission or information storage and retrieval, electronic adaptation, computer software, or by similar or dissimilar methodology now known or hereafter developed.
The use of general descriptive names, registered names, trademarks, service marks, etc. in this publication does not imply, even in the absence of a specific statement, that such names are exempt from the relevant protective laws and regulations and therefore free for general use.
The publisher, the authors and the editors are safe to assume that the advice and information in this book are believed to be true and accurate at the date of publication. Neither the publisher nor the authors or the editors give a warranty, expressed or implied, with respect to the material contained herein or for any errors or omissions that may have been made. The publisher remains neutral with regard to jurisdictional claims in published maps and institutional affiliations.

Cover illustration: © Alex Linch shutterstock.com

This Palgrave Macmillan imprint is published by the registered company Springer Nature Switzerland AG
The registered company address is: Gewerbestrasse 11, 6330 Cham, Switzerland

Naomi and Jude

Foreword

Election fraud and other voting anomalies encompass technological, social justice, and systemic issues are rampant throughout the world, including in the USA. I voted in Orange County, Florida, in the infamous 2000 Presidential election that pitted George W. Bush against Al Gore. The election, my first in the State of Florida, turned out to be one of the most controversial in US history. *Hanging chads*[1] entered the everyday vernacular and quickly became national news. The sheer number of hanging chads generated by Florida's faulty voting machines meant that many votes cast in good faith were not registered properly, leading to an Electoral College margin that was "so close that it took one's breath away" (Elving, 2018).

> As the Electoral College vote took shape on election night, with the results piling up from around the country, it was clear the vote in Florida was going to determine not only the winner of that state's 25 electoral votes but the next occupant of the Oval Office. Although Gore had won the popular vote by roughly a half-million ballots, the all-important Electoral College count from the other 49 states (and District of Columbia) was so close that whoever won Florida would be the overall winner. (Elving, 2018)

---

[1] A *chad* is a tiny bit of paper that is punched from a ballot using a punch-type mechanical voting machine. A hanging chad is one that is not fully separated from the ballot during voting.

The election wasn't ultimately decided until the Supreme Court ruled in favor of Bush on December 12, 2000. And by some accounts, including the following summary by the nonpartisan voter advocacy organization FactCheck.org, the 2000 Presidential Election results are still in dispute.

> According to a massive months-long study commissioned by eight news organizations in 2001, George W. Bush probably still would have won even if the U.S. Supreme Court had allowed a limited statewide recount to go forward as ordered by Florida's highest court.
>
> Bush also probably would have won had the state conducted the limited recount of only four heavily Democratic counties that Al Gore asked for, the study found.
>
> On the other hand, the study also found that Gore probably would have won, by a range of 42 to 171 votes out of 6 million cast, had there been a broad recount of all disputed ballots statewide. However, Gore never asked for such a recount. The Florida Supreme Court ordered only a recount of so-called "undervotes," about 62,000 ballots where voting machines didn't detect any vote for a presidential candidate.
>
> None of these findings are certain. (Jackson, 2008)

The recount and delay of election results, though supremely disruptive to the country as a whole, was not even Florida's only election upheaval that year. A comprehensive report by the US Commission on Civil Rights revealed eight distinct areas of voting violations in the 2000 election, including, but not limited to, "allegations that Florida voters were prevented from casting ballots or that their ballots were not counted" as well as "allegations of widespread voter disenfranchisement in Florida." The Commission is authorized—and obligated—to investigate all claims that suggest "any pattern or practice of fraud" and any infringement on the right of citizens "to vote and have votes counted."

As this brief trip down memory lane suggests, election fraud and other voting anomalies can be blamed on technological, social justice, and systemic issues, among others. Voting controversies have not disappeared from US politics in the nearly two decades since the fraught 2000 Presidential election. If anything, they are more visible than ever, despite innovations such as biometric verification. Moreover, the US is far from alone in its struggle against election fraud and voter disenfranchise; as Isidore Dorpenyo's unique research illustrates, election fraud is a worldwide problem with a complex array of potential—if often partial—solutions.

Dorpenyo inserts the democratic practice of electing public officials squarely into the conversation surrounding international technical communication scholarship and practice. *User Localization Strategies in the Face of Technological Breakdown: Biometric in Ghana's Elections* is, of course, set in Ghana, Dorpenyo's native country, a post-colonial democracy in West Africa. Government corruption and inconsistent record-keeping have allowed an epidemic of over-voting to take place. People vote more than once; unregistered people (both citizens and non-citizens) cast ballots illicitly; even minors manage to vote. In addition, according to the Trading Economics website, around 45% of Ghanaians live in rural areas, so uneven access to polls in remote locations may exacerbate inequities. In 2012, in an attempt to combat what they saw as rampant voting fraud, the government of Ghana decided to adopt biometric authentication, defined by security firm Gemalto as a "security process that relies on the unique biological characteristics of an individual to verify" his or her identity. Biometrics are commonly used by law enforcement, border security personnel, health identification, and, as in Ghana and elsewhere, for voter registration and other civil identity applications. Broadly, their purpose is "to manage access to physical and digital resources such as buildings, rooms and computing devices" (Gemalto, 2019).

Dorpenyo employs stakeholder interviews and genre analysis of marketing materials and instructional documentation to closely examine the government of Ghana's process of implementing the biometric verification device (BVD) for voter registration and authentication. Operating with a decolonial stance and a technical communication scholar's lens, he augments the strategy of technological localization that Nancy Hoft introduced to the field of technical communication more than 20 years ago (Hoft, 1995), melding it with Johnson's (1998) user-centered design framework and Sun's (2012) attention to the rift between designers' and users' cultures. Dorpenyo's detailed, well-researched, and carefully contextualized longitudinal study, while providing a social justice perspective on enfranchisement, culminates in a set of best practices for technical communication researchers, teachers, students, and practitioners who are engaged—as most of us ultimately are—in international and intercultural technology transfer. For example, he arrives at three localization strategies: linguistic localization, user-heuristic experience localization, and subversive localization, which operate within what he calls a localization cycle. Each of these manifests somewhat differently with different outcomes and distinct affordances and constraints.

Dorpenyo's stake in the proper conduct of elections may have begun when, as a child, he helped his father run for public office. But his professional affiliation with the field of technical communication leads him to this thoroughly researched case, which complicates and interrogates the transfer of "Global North" technology to the "Global South" as much more than an instrumental process.

Houghton, MI, USA

Karla Saari Kitalong, Ph.D.
Professor of Humanities
Michigan Technological University

References

Elving, R. (2018). *The florida recount of 2000: A nightmare that goes on haunting.* Retrieved from https://www.npr.org/2018/11/12/666812854/the-florida-recount-of-2000-a-nightmare-that-goes-on-haunting.

Gemalto. (2019). *Biometrics: Authentication and identification.* Retrieved from https://www.gemalto.com/govt/inspired/biometrics.

Hoft, N. L. (1995). *International technical communication: How to export information about high technology.* Hoboken, NJ: Wiley.

Jackson, B. (2008). *The Florida recount of 2000.* Retrieved from https://www.factcheck.org/2008/01/the-florida-recount-of-2000/.

Johnson, R. (1998). *User centered technology: A rhetorical theory for computers and other mundane artifacts.* Albany: State University of New York.

Sun, H. (2012). *Cross-cultural technology design: Creating culture-sensitive technology for local users.* New York: Oxford University Press.

ACKNOWLEDGEMENTS

A book is never the work of a single author. Rather, it is an articulation and re-articulation of ideas from a network of people. I am fortunate to have had such support come my way from a variety of people I met along the way when working on this book. I owe these people my every gratitude. The first of such persons is Karla Saari-Kitalong whom I have come to identify as my post-dissertation mentor. Thank you for inspiring me to undertake this project and for guiding me to think critically about every aspect of the project, and the argument I make in this book. Right after my dissertation, you helped to organize think about how to switch a project from dissertation mode to a book mode. You were there for me from the very first day I started this book to the last day. Even though you were busy, you took the time to read every chapter and offer insightful feedback. In addition, Shelley Reid and Godwin Agboka read my manuscript and offered extremely insightful criticisms for the book. Shelley, I cannot forget the hours we spent in your office trying to map out the contours of the project and how to better foreground my argument and make it relevant to the field of technical communication. It came out well and it helped me to focus on localization. Karla, Shelley, and Godwin helped me to sharpen my argument and analysis. Kirk St. Amant also read chapters and offered wonderful guidelines. He helped me to identify presses I could send my work to. Thanks a lot!

I am also indebted to my dissertation committee members, Ann Brady, Robert Johnson, Marika Seigel, and Godwin Agboka. To Ann Brady, my supervisor, and number one cheerleader, I say I am eternally grateful. You never stopped making me believe that my work is "top notch." Your excitement about my project kept me going. Also, your feedback, weekly meetings, and guidance have brought me this far. I also appreciate the efforts of Robert Johnson, Marika Seigel, and Godwin Agboka for sharing their depth of knowledge, insights, and advice to me. Their final comments strengthened the structure and content of my work. My committee members are awesome!

I also benefitted immensely from conversations with colleagues at Michigan Technological University and at Conferences: Laura Gonzales, Akwasi Duah-Gyamfi, Keshab Acharya, Jessica Lauer, Amanda Girard, Joana Schreiber, Valorie Trosch, Ruby Pappoe, and Regina Baiden. My good friend Laura Gonzales, one of my reviewers, offered invaluable feedback. I am also grateful to my colleagues at Mason: Debra Lattanzi Shutika (my Chair), Douglas Eyman, Shelley Reid, Heidi Lawrence, Alex Monea, and Steve Holmes for their support.

My initial data collection benefitted from a grant from the Humanities Department at Michigan Tech, and the second phase of data collection was supported by a research start-up fund from the English Department at George Mason.

I was lucky to have Rachel Daniel as my editor. Thank you Rachel for your patience and wise counsel. Madison Allum, Rachel's assistant, was also helpful in this process. My reviewers offered invaluable feedback on my work.

Electoral Commission officials across the country were generous with their time and resources. Thanks to all those who willingly participated in this project.

Finally, I thank my family and friends for their endless support. My friend, Emmanuel Agyapong and his wife Linda, gave me a place to sleep when I travelled to the Western Region to interview participants. Emmanuel also went to the extent of looking for participants to be interviewed. I benefitted from his friendship greatly. My cousin, and childhood friend, John Konoh Tordzro, who also happens to be a volunteer for the Electoral Commission, helped me to find participants in Accra. I couldn't have made it without his support and dedication. I have also enjoyed the support and encouragement of Philomena Yeboah, Ellis Adjei Adams, and Joyce Yenupini Adams. My in-laws, Ebenezer Appiah and Martha Plange, gave me emotional support. So did my parents, Joana Afiadenyo and Francis Dorpenyoh, and my siblings, Tarcisius Edem Dorpenyo and Irene Sena Dorpenyo. My Wife, Naomi Appiah, and my nine-month-old baby, Jude Selikem Dorpenyo, filled me with enough energy and support throughout the process. Jude taught me that time management was of the essence if I could achieve anything. I cherished every moment I spent joggling between diaper changing, feeding, lulling Jude to sleep, and writing a monograph. I have also benefitted enormously from the Kufour and Minyila families. Occasionally, the Minyila's took the burden off my shoulders by caring for little Jude. Our casual conversations lightened me up and gave me enough energy to pursue this project.

Contents

1. Recovering the Lost Voices of Users in Localization — 1
2. Biometric Technology: The Savior of a Risky Electoral System — 37
3. Decolonial Methodology as a Framework for Localization and Social Justice Study in Resource-Mismanaged Context — 53
4. Stories of Users' Experiences — 79
5. Linguistic Localization: Constructing Local/Global Knowledge of Biometric Technology — 91
6. User-Heuristic Experience Localization — 129
7. Subversive Localization — 145
8. You Are Not Who You Say You Are: Discriminations Inherent in Biometric Design — 185

9 Conclusion: Participatory User Localization   201

Bibliography   221

Index   233

List of Figures

Fig. 3.1	Emergent categories derived from grounded theory analysis of interview data	74
Fig. 5.1	Educational material defining biometric technology	97
Fig. 5.2	Similarities between old and new registration systems	98
Fig. 5.3	General advice to voters	99
Fig. 5.4	Oblique overview of BVD	101
Fig. 5.5	Instructions showing outcomes of verification	102
Fig. 7.1	Functional elements of the biometric device	153
Fig. 7.2	The process of unpacking the biometric device	154
Fig. 7.3	Message indicating when to clean the scanner	155
Fig. 7.4	Message indicating biometric should be handled with care	156
Fig. 7.5	Laser beam	157
Fig. 7.6	Instructions on how to switch the device on/off	160
Fig. 7.7	Instructions on how to scan the barcodes on register with biometric barcode	161
Fig. 7.8	Instructions cautioning on staring at laser in the barcode	162
Fig. 7.9	Linear model of the user manual design process	165
Fig. 7.10	The oblique picture of the biometric device used during the elections	176
Fig. 7.11	Instructions showing possible outcomes of verification	177
Fig. 7.12	The voters register in view	180
Fig. 7.13	Verification procedures for verification officers	181
Fig. 9.1	Localization cycle	203

CHAPTER 1

Recovering the Lost Voices of Users in Localization

As a young first child of my parents, I had the opportunity of following my father to political campaigns. Yes, I remember vividly how I enjoyed carrying his food and water along anytime he went on political campaigns. I also recall the many times I have enjoyed chants of political songs and the sense of comradery exhibited by party members during campaigns. Growing up in an environment that was constantly filled with political conversations and party paraphernalia, I had hoped I was going to become the son of a Member of Parliament. Unfortunately, this burning desire to become the son of a Member of Parliament, mostly because of the social prestige that came with the position, never materialized because my father could not secure the required votes needed to beat the parliamentary candidate of the National Democratic Congress (NDC), one of the two main political parties in Ghana.

I recollect with much clarity the pain that came with both defeats. Those were moments when the entire family could lugubriously go days without a shower; there was less appetite for food and no desire to even switch on the radio or television in earnest desire to avoid listening to election results or just to avoid hearing about the success story of the opponent. Our nostalgic reminiscences of political campaigns were painful to think about: the several moments we walked for miles to places which had no accessible roads; the days we left home very early and came back the following day; moments when we abandoned our campaign cars because they got stuck in mud; and several instances where we had to put up with people who verbally abused us. In both of his losses, my

© The Author(s) 2020
I. K. Dorpenyo, *User Localization Strategies in the Face of Technological Breakdown*, https://doi.org/10.1007/978-3-030-26399-7_1

father was persuaded by his followers to reject the election results. These people cited instances of vote rigging, ballot box snatching, harassment of polling officers by opposing party, impersonation, and voting by minors. Against all odds, my father conceded defeat. At that young age, I said to myself that if instances cited by father's followers were anything to go by, then something needed to be done.

Frustrated by constant conversations about these alleged electoral malpractices, I said I was never going to vote again. Suddenly, the Electoral Commission of Ghana (EC) announced that it was going to adopt a biometric verification device (BVD) to enhance the country's electoral process. According to the EC of the country, the biometric was going to: assist in detecting and preventing practices of impersonation and multiple voting; expose electoral offences; provide transparency in results, and make it very hard for someone to use the particulars of a different person to vote. With the representation of the biometric in such a positive light, many Ghanaians went to the polls with hopes, but little did the election management body conceive that the biometric would introduce new challenges into the electoral process.

Now imagine that you confidently walk to the polling station with hopes that you are going to register or vote only to realize that when you put your fingers on the biometric technology for authentication or verification, the biometric fails to recognize your fingers. You try again and it fails to pick your fingers. You try for the third time, but the technology indicates that you are not who you say you are. Frustrated with the technology, you give up. Which means you cannot vote. On a scarier note, imagine that you go to the polling center to vote only to realize that the only biometric technology in the voting center has broken down or batteries of the technology were constantly draining because the technology performs poorly under dusty, hot, or humid weather conditions. Or, that the biometric has broken down because election officials did not obey instructional procedures. As if these breakdowns were not enough, the printers used during the elections also started breaking down because they could not take the pressure. The consequence of these breakdowns or rejections was that people were disenfranchised. An EC official I interviewed, for instance, indicated that:

> then come election day it broke down, some people couldn't use it, some people had to use the manual registration which was outside the law and in fact some people got disenfranchised because the machines broke down

and when it was rescheduled not all people were able to come back so these were the initial problems with the use of the device…When you take the verification device, for example, the printers were just breaking down like that because they could not take the pressure. If you start printing, you print 1, 2, 3, 4 then the printer breaks down… the BVD failed and Superlock Technologies Limited (STL), the technicians, also blamed it on humidity, high temperature.

If you happen to be near or in the center of this scene, how would you feel? These anecdotes indicate that the biometric technology broke down on several levels: (1) biometric performed poorly because it could not withstand the heat in Ghana; (2) the machine could not read the fingerprints of some voters; (3) training in biometric use didn't really help since most users struggled to use the technology on election day; and (4) user instruction manual was confusing. Realizing the severity of the problem, the EC and voters started adopting local measures to salvage the situation: Those who were rejected were asked to use Coca-Cola, local herbs, and detergents such as OMO to wash their hands, and canopies were used in some polling stations to control the temperature.

The biometric breakdowns in Ghana reveal that designing for global use is challenging. Designing for global users means thinking about the broader context within which a product or technology will be used. Broader context, as I use in this book, acknowledges a relationship between weather conditions (or physical environment), the space, location and place of technology use, the users of the technology, how the technology will be used, what situation will trigger the adoption and use of the technology, the needs of the users, and when it will be used. This means there is a need to understand that "context is not about a superficial interaction. It's about deep engagement [with] and an immersion in the realities and the complexities of our context" (Douglas, 2017). Thinking about and engaging in these broad contextual issues have proven to be daunting tasks for designers, because in most cases the designers of technologies we use do not even know which user will purchase their products and how those users will even put the technology to use. In the same way, in most cases, users do not know the designers of the technology they purchase and use. For instance, the EC officials of Ghana did not have any knowledge of the company which designed the biometric technology in use.

It is thus an established fact that a designer may never meet or know about users of the technologies they design. This bitter truth is tacitly expressed by Jonathan Colman, a experienced product user and content strategist, when he revealed one of several 'wicked ambiguities' UX officials encounter, "the challenges of creating solutions for people whom we'll never know in our lifetime" (Colman, 2015). Even though Colman is addressing user experience (UX) experts, this is the reality most designers will grapple with for a lifetime: Users will only be represented with mental models. Huatong Sun makes this more revealing when she identifies that a gap exists between the product designer and the product user. This gap is as a result of the existence of two levels of localization: localization at the developer's site and localization at the user's site (Sun, 2004, p. 2). This gap, as I see it, presents one underlying issue: the clash of cultures. In one instance, there is the culture of design which influences how a technology should be designed and used, and on the other, there is the culture of use. The disconnect between culture of design and culture of use can result in the mass breakdown of technology as was the case in Ghana.

Culture as a Problem and a Relevant Factor in Cross-Cultural Design Practices

Scholarship indicates that designers are aware of the wicked ambiguity Colman hints at, so they work hard to resolve the tension that exists between the culture of design and user culture either by internationalizing, localizing, or customizing their products to make them appeal to global users (Esselink, 2000; Hoft, 1995; Sun, 2012; Taylor, 1992). Interestingly, these processes used by product designers emphasize the importance of "culture" in producing globally acceptable products. Taylor (1992), for instance, reveals that internationalization occurs because developers seek to extract "the cultural context from a package" (p. 29). The end goal of this process is "to be able to have a sort of generic package, with an appendix or attachment that details all the culturally specific items" (p. 29). It is obvious from this that designers extract culture from their products because they have become increasingly aware that the local culture in which they design products shape the way their products are designed and used. As a result, they find it appropriate to try to move beyond their local culture to make the product appealing to other cultures. Localization, on the other hand, is defined by Dave Taylor as "taking something that is designed for the international market and adding features and elements to better match the target culture and marketplace" (p. 29). In this process, a parent company designs a product and sends to a localization company which then is responsible for appropriately designing local versions of the buttons, controls, packaging, documentation, etc. Localization takes the generic product and transforms it into a one that matches the cultural expectations of users. The localization firm becomes the "local agent" which works to fit the product into specific local rhetorical cultural values.

Similarly, the Localization Industry Standards Association (LISA) defines internationalization as "the process of generalizing a product so that it can handle multiple languages and cultural conventions without the need for redesign" (cited in Esselink, 2000, p. 2); and localization as the process of "taking a product and making it linguistically and culturally appropriate to the target locale (country/region and language) where it will be used and sold" (cited in Esselink, 2000, p. 3). While internationalization is the first stage which prepares a platform for cultural and linguistic features embedded in a product to be "extracted and generalized, localization is the completion stage, where the product is fine tuned for the specific market niche that is targeted" (Taylor, 1992, p. 33). Dave Taylor also adds that in localizing a product, designers realize that "all elements of a particular culture or society must be viewed from the point of view of that culture, rather than that of the viewer" (p. 35). Some of these broader cultural elements include transliteration, hyphenation, spelling, collation, notational conventions, numbers, currency, time, and date.

Indisputably, internationalization and localization have instrumental benefits as these practices increase sales and also reduce the level of resistance from target cultures; the downside is that many approaches to internationalizing, localizing, or customizing focus on how designers make changes to meet local cultural needs. For me, these approaches to resolving the design gap are unidirectional, meaning that these solutions only come from the culture of design, represented by engineers whose emphasis is on functionality, and little to no input from the culture of use. Even though localization takes into cognizance broader cultural context of use, the extraction of culture embedded in products are nevertheless affected by designers or localization agencies. This reflects a top-down approach to product design or it "raises the danger of universalizing and othering users" (Agboka, 2013, p. 41).

Recent work in technical communication calls this approach to localization to question, arguing that current discussions on localization pose a problem mainly because the concept has been narrowly defined. Specifically, they argue that localization suffers from a narrow and static definition of culture (Agboka, 2013; Sun, 2004, 2006, 2009b, 2012). This is because methods used to collect data about culture only capture dominant or large cultural characteristics to the neglect of use activities in a locale. Hoft (1995), for example, proposes two methods international researchers can use to collect data: the Iceberg model and the International Variables Worksheet. The former helps individual researchers to focus on the obvious characteristics of a culture (which is the part of the iceberg above the water level) and work their way down to the unspoken and unconscious rules (these are below the water surface). The obvious characteristics only form about 10% of the culture, while the other traits submerged under water form about 90% of cultural traits (p. 59). The international variables, on the other hand, help the research to document political, economic, social, religious, educational, linguistic, and technological characteristics of the country. More so, one of the most used approaches proposed by Hofstede emphasizes the importance of focusing on such traits as power distance, uncertainty avoidance, individualism and collectivism, masculinity and femininity, and long-term orientation versus short-term orientation (Hoft, 1995, pp. 84–88). The implication of these approaches is that when you enter a community, the obvious thing you capture is language, the food people eat, the sport they value, time, date, and perhaps how they relate with one another.

The consequence of this monolithic approach to capturing culture is "poor user experience" (Sun, 2012, p. 5) because the frameworks capture culture in abstract terms while also separating culture from use situations in a localization process (Sun, 2012, p. 13). More worrying, the action of users is missing because little effort is put in place to study users. In essence, users have not been cast as agents of change. Rather, users have been "constructed as passive consumers … with little or no agency to create and re-create…" (Agboka, 2013, p. 30). Therefore, previous scholars unanimously call for a definition of localization which emphasizes and centers on the user. Agboka (2013), for example, proposes that we reconfigure localization "as a user-driven approach, in which a user (an individual or the local community) identifies a need and works with the designer or developer to develop a mutually beneficial product that mirrors the sociocultural, economic, linguistic, and legal needs of the user" (p. 44), and the core of Sun's scholarly works contend that localization should lead to an understanding of user activities in context. She makes us understand that a gap exists between the product designer and the user of the product. The tension between localization on the two levels should lead to the development of "an effective way to address cultural issues in IT localization and design well-developed products to support complex activities in a concrete context" (Sun, 2004, p. 2). She also identifies that the lack of a broader understanding of culture hurts localization practices because localization specialists only focus on delivery aspects, that is, "specialists only pay attention to such issues as colors, page layouts, or how dialogue boxes should be resized for a certain language" (Sun, 2006, p. 460), and this leads to a neglect of the broader sociocultural contexts where "information products are situated, and where products are designed, produced, distributed and consumed" (Sun, 2004, p. 2).

This book extends the conversation about localization championed by technical communication scholars; it argues that even though localization has economic values to companies, it must also magnify the agency of users because the success of a technology depends on how it meets user needs. Success can also depend on the creative efforts users put into use situations. For this to happen, our gaze should not only be fixed on understanding the cultural relevance of localization, but also the rhetorical nature of localization. By this, I imply that localization should be about the relationship between rhetoric, that is, the powerful use of language to articulate the powers of a technology and culture. In Ghana's case, for example, I focus on the strategic use of language to represent the biometric as the extraordinary technology that can "expose" electoral offenders, "provide transparency" in results, "prevent incidence of multiple voting" or "make it extremely difficult for a person to use the name and particulars of another voter." These words and phrases used, I contend, shaped the beliefs and expectations of some Ghanaians about the biometric technology and the 2012 elections in general. The biometric technology is articulated as the "savior" or "the solution" to the electoral woes the country has experienced over the years. This can include "both verbal and visual discourse, both public and interpersonal communication and both explicit and implicit arguments" (Scott, 2003, p. 3). My line of argument, thus, conceives of localization, as the extent to which users demonstrate their knowledge of use by adopting and reconfiguring the purpose of technology to solve local problems.

My definition of localization is an extension of Dourish's (2003) use of the term "appropriation." For me, localization and appropriation can be used synonymously to mean two things: first, the adoption, adaption, and incorporation of technology to meet local exigence, and second the extent to which a technology is used for "purposes beyond those for which it was originally designed, or to serve new ends" (Dourish, 2003, p. 467). Hence, localization is about subversion, in the sense that users are able to reconfigure and subvert the intended use of technologies designed and not about translation (Gonzales, 2018), because translation has the proclivity to focus on attempts made by users to "replicate the meaning of a word from one language to another" (Gonzales & Zantjer, 2015, p. 273). The implication is that translation, as a form of localization, only pays attention to language use, but localization should be beyond the focus on language. At the very least, we can argue that when users are able to subvert and reconfigure the original intent of a technology, they enact some form of agency and "knowing." Sun (2012), as an example, refers to this use of technology for purposes beyond their intended use "local use" and argues for why user interpretation should be taken seriously. Localization helps users to talk back and re-right definitions that cast them as knowledgeless. As I will indicate in later chapters, "talking back" and "re-righting" are championed by decolonial projects such as what I advance in this book. I argue that local use determines how the intended purpose of a technology is subverted by users. Therefore, local use and subversion play an important role in the localization processes I advance in this book.

Ghana's biometric use exemplifies these two definitions. First, realizing that no local method could resolve deep-seated electoral irregularities, the EC sought a more appropriate technology which has a reputation for identifying miscreants. In this regard, the biometric is the appropriate technology since the technology could capture and identify individuals accurately. And second, by using the biometric to enhance elections, the country in a way reconfigured the purpose of the biometric technology. This is also an indication that the setting within which a biometric can be used is not fixed; it is mutable, contingent, and flexible. To be clear, biometric is designed to be used primarily as a security apparatus and not as an election technology. Thus, it is normal to encounter biometric technologies at security posts such as airports, prisons, and banks. It is because of the reconfiguration of the purpose of the biometric that I find it appropriate to discuss and advance conversations about localization. By using the biometric for purposes beyond its original intent, Ghanaians indicate that there are multiple roles the biometric technology can perform and that designers should think about what else to include in the design of the technology. For instance, Ghana's electoral system suffers from some malpractices such as foreigners and minors voting, but the biometric is not able to detect these anomalies.

Nonetheless, a localization focus is relevant for four reasons: First, it is an indication that individuals who are far removed from the design of a technology can repurpose and use a technology to resolve their local problem; second, it is an avenue to get designers of products to understand that localization does not end immediately a product is shipped from their production companies; third, it is necessary to move beyond functionality to focus on the social world in which a technology will be used. Finally, localization exposes the challenges users go through in integrating technology to fit into their local systems. Ghana's use of the biometric technology is a single case, but it indicates to designers that when a technology leaves its manufacturing facility, designers have no major control over how that technology is used. To this end, it will be relevant for designers to see design as a process which continues outside of design and localization as the medium through which designs can accomplish different needs and purposes of users. This means designers should (1) instill flexibility into their design, and (2) see localization as a collaborative effort between users and designers and not necessarily between product designers and localization companies. This is because users may have different uses for their products and also because users have knowledge of the complex use situations in which technology will be situated.

The numerous breakdowns indicate that the EC officials, and perhaps designers, were interested in communicating the "operational and instrumental affordances" to the neglect of "social affordances," that is, "the properties of a technology that support object-oriented activity and social behaviors in a sociocultural and historical context" (Sun, 2006, p. 460). As Sun indicates, the consequence of the neglect of sociocultural factors in design is that a gap is created between designer and user. I see the study of the biometric breakdowns as a continuation of Huatong Sun's argument that challenges abound in creating technology for users from a different culture, and thus it is necessary to attend to action and meaning cross-cultural design. Interestingly, when the biometric technology broke down, EC officials devised several tactics to salvage the situation—thus making user tactics/strategies a core part of conversations about localization. The various strategies used indicate that users do not wait for designers to save them when technology breaks down. They save themselves. As I will indicate later, the actions enacted by individual EC officials and voters came to represent "institutional strategies" (Kimball, 2017, p. 4) for resolving the biometric breakdowns as different polling stations repeated the actions to resolve similar biometric failures. One polling station uses a locally manufactured detergent known as OMO to wash the hands of the voter who was rejected; another polling station hears of it and adopts the same strategy and eventually it becomes a solution to the problem. Or one polling station uses Coca-Cola to clean the hands of a rejected voter; another polling station adopts it and it paves the way for others to emulate.

My intention in this book, therefore, is to examine how the adoption and use of biometric in Ghana's 2012 and 2016 elections help to advance knowledge on how users strategize to localize technologies to fit their local needs and conditions and what such strategies reveal of users. In Ghana's case, I identify and advance three localization processes: linguistic localization, subversive localization, and user-heuristic localization. I seek to foreground the expertise of local users by examining how they adapt and use an unknown technology to solve local problems that have bedeviled Ghana's electoral system. I believe an examination of users' experiences will expand on Sun's (2006) idea that localization exists on two levels—the developer level and the user level—and Agboka's (2013) call for the active participation of users in localization. While I am engaged in advancing conversations about localization, I will also reveal the complex entanglements surrounding the use of the technology, how it is socially constructed, how the biometric becomes a tool for racial discrimination, and how it even becomes a site of struggle in a local context.

This line of argument I pursue is emphatic about the user and the situation of use. The articulated user is not passive, he/she becomes, as Johnson (1998) identifies, "an active participant who can negotiate technology use" (p. 33). Johnson captures the form of active negotiation of users in three concentric circles. The first ring captures the activity of the user as he/she engages in *learning*, *doing*, and *producing* the technology. These three models, he states, "define global activities that users are apt to be involved with during either the design, dissemination, or end use of technological systems or artifacts" (p. 38); the next outer ring depicted as *disciplines, institutions,* and *communities* "emphasize constraints that larger human networks place upon technological use" (ibid.). In addition, the rhetorical model considers the importance that *culture* and *history* play in technology use. As Johnson avers,

> Cultural forces define nearly every human action, and in a world more dependent than ever on international communication and technology transfer the element of culture is without question essential when defining the use of a technology. History... refers to the reflective aspect of understanding human action,...Thus history informs our understanding of technology in unique and indispensable ways. (p. 39)

Lumped together, Johnson helps us to understand that the complex context of use serves a plethora of purposes: "it serves as a heuristic for analyzing technological artifacts or processes. It also can be a mode for exploring the people who use, make, and/or even destroy technology. It can help tell tales of people as they struggle with, become enamored of...technology." Finally, technical communicators can use the complex of technology use "to study the audience we refer to as users" (p. 40). This rhetorical model becomes relevant to my project in the sense that it stresses the roles that disciplines, institutions, sociocultural and environmental factors, and communities play in technology use. More so, it emphasizes the importance of culture and history and how these two forces shape technology adoption and use. In entangling these broader situational issues, I keep focussing on revealing user strategies.

THE USER IN LOCALIZATION AND USER-CENTERED APPROACHES

About two decades ago, Robert Johnson shifted our attention from the study of technology developers and designers to users of technology. In his book *User-Centered Technology*, he reveals that the user occupies the voiceless position in the matrix of technological system; the users' voice is undercut by conversations that cast the designer as the expert of technology and the domain of knowledge making about technology. The user, on the other hand, dwells in the "land of the mundane" (p. 3) and, thus, is seen as that person who sits at the "end of the technology development cycle who take[s] a tool, use[s] it to make something, and then go[es] on from there" (p. 9). Contrary to this narrative, Johnson advances and recovers the voice of technology users by arguing that knowledge does not only lie in the hands of the developer but it equally lies in the head of the user. The user knows something about technology; the user has "knowledge of use" (p. 13). Through a rhetorical lens, he hands over power to the user of technology and called on fellow technical communicators to research into the various ways users of technology exhibit knowledge of use. This call has been received tremendously. On the one hand, a group of scholars advocate that users should be part of the design of technology (Brady, 2004; Johnson, 1998; Jones, 2016; Salvo, 2001; Seigel, 2013; Walton, 2016). On the other hand, other scholars are advocating the study of UX of technology in order to communicate user needs to designers (Sun, 2006, 2012). More recently, still, other scholars are advancing a human-centered approach to technology design (Jones, 2016; Rose, 2016).

While these conversations are encouraging, there remains one problem that needs to be resolved: Little has been done to account for user needs in unenfranchised/disenfranchised cultural sites or international/cross-cultural audiences (Agboka, 2013; Sun, 2009b, 2012), or not much is done to highlight the expertise of users' who adopt and use technology to enhance their daily lives in resource-constrained societies. In most cases, users are cast as victims of technology. To be sure, users from such areas have not been represented as "agents who initiate and implement change themselves" (Agboka, 2013, p. 30). Agboka advises for users to assume the position of knowledge and agency: (1) localization should happen at the user's site, where prevailing local conditions influence design; (2) localization needs to be a collaborative effort between user and developer; (3) localization should not be looked at in terms of translation; (4) localization should be based on long engagement in the target community; (5) localization should pride itself on relevance of local needs; and (6) localization should be based on mutual understanding (2013, p. 45). Similarly, in an attempt to identify how technologies are designed to meet the needs of cross-cultural users, Sun (2012) identifies that technology design processes create "a disconnect between action and meaning" for local users (p. 8). That is, little is done to incorporate the meanings, behaviors, practices, and state of consciousness of users in cross-cultural contexts into technology design. I have found no better description of this very point than in the words expressed by Mr. Kofi Manu,[1] one of the EC officials I interviewed: "…if the commission is actually part of it [design process], we will ask 'will this equipment withstand heat, cold or whatever it is.'"

Undoubtedly, the statement by the EC official reflects the voice of a user who is frustrated by a technology which has been adopted for use and is expressing the sentiment of many others: We want to be part of decision-making about technology design. The EC official reinvigorates Johnson's idea that the user has knowledge of use. For instance, Mr. Manu reveals that most polling stations in Ghana are in the open:

> if you look at most polling station set up, they are in the open [you know] ah ha some you can get them under trees, under the corridors of certain buildings not in the buildings…Assuming that all those polling stations were to have shades or canopies, we could have actually avoided some of those issues because we know that it [biometric device] cannot work well under scorching sun,

and Mr. Alfred Gadugah, another EC official I interviewed, exclaimed

> … All that I can say is that those who produce these machines for us should take cognizance of our peculiar environment and then find ways and means of ensuring that they are able to withstand the vagaries of our weather here You see, WE VOTE IN DECEMBER, harmattan and when you ever go to Western region or the north you see the dust coloring the whole skyline. The whole skyline becomes yellow, brownish yellow because of the dust so in some of the areas you can't avoid the dust in December when you are voting.

In these extracts, the officials are aware that the weather or the environment affects technology use. It is possible to say that the neglect of contextual factors, such as the space within which the biometric was going to function, the high temperatures in Ghana, could have led to the breakdown of the biometric technologies in the country. The revelation by EC officials that the biometric used in most cases were constantly overheating and consequently breaking down in most polling stations is worthy of note. Perhaps, this recast the idea that the EC official has local knowledge of use, something the designer of the technology lacks. This scenario indicates that users know their "local culture and contexts better than" designers and even though they may "not be able to articulate those cultural and contextual factors well…they know what works in their own contexts…" (Sun, 2006, p. 458). It is even possible to say that the design features of the biometric reinforce the Western idea that climate control (Air conditioner and heat) will be available to maintain consistent and technology-friendly temperature.

Victor van Reijswould and Arjan de Jager (2011) argue profoundly that environmental factors affect the design and transferability of technologies to other parts of the world. In most cases, these technologies come with assumptions from the designers which may not be helpful to the local communities that adopt them. They state, for instance, that "hardware is not protected against the dust, sand, and heat of African countries" (p. 74). Tania Douglas extends the conversation by van Reijswould and de Janger by revealing how medical equipment imported to Africa, for instance, do not function properly because too often they "are not suitable for local conditions. They may require trained staff that aren't available to operate, maintain and repair them; they may not be able to withstand high temperature and humidity and they usually require a constant reliable electricity." She concludes thusly: Africa has become the "equipment graveyard," that is, it has become a "typical final resting place for medical equipment from hospitals in Africa" (Douglas, 2017). The same could be said about technologies adopted to enhance other sectors of African economies.

The lack of attention to weather and environmental factors and the skill of EC officials are perhaps some of the fundamental reasons why Ghana experienced breakdowns of the biometric technology during the elections. For example, the "tips" section in the user manual that accompanied the biometric device used warned that: "...the device must be operated in a shady area devoid of *direct sunlight and heat*," and "Keep the device away from liquids, wet surfaces and *dust*" (manual, p. 5). Mr. Gadugah thinks this warning note is ludicrous because as he indicates:

> you know we vote in the open. We have our polling stations on corridors, under canopies and when you keep telling us that we shouldn't put it in a direct sunshine, do you think somebody will in his right frame of mind, carry this thing and go and sit in the direct sunshine? So what exactly do you mean by we shouldn't be in direct sunshine, okay? We are under canopies, we are under trees and we are on corridors so I don't know at which point we were supposed to be in a direct sunshine or shady always we are in a shady area but mind you if you ever sat under these canopies, the heat that the canopies radiate is even more than going directly under the sunshine.

These expressions help us to think about how much thought was put into how weather conditions will affect the use of the biometric device adopted. Geographically, Ghana lies between latitude 0° N and 23° N. This means that the country lies a little above the equator and within the tropical zone. The implication of this geographic location is that the country is relatively warm or hot all year round. In addition to the warm conditions, Ghana also experiences a dry weather condition between November and March, a period which is usually referred to as the Harmattan. During this time of the year, dry winds blow from the Sahara Desert across the country. The implication is that the country experiences dusty, humid, and hazy conditions. It is not any wonder that "the machines faced technical hitches that led to the malfunction in some polling stations…" (*Report of the Commonwealth Observer Group: Ghana Presidential and Parliamentary Elections*, 2013, p. 18). I assume that a user who was part of the design process could have informed designers about the setup of polling stations and how best to design to meet local weather conditions. These are indications that use and user situations have been narrowly defined by both designers and users.

Moving Beyond the Narrow Representation of Users

Because users have been neglected from technological development processes, and because technology use is situated in a complex context, users have to adopt strategies to localize technologies to fit their everyday lives. This interest in user strategies is well explored in Huatong Sun's book *Cross-Cultural Technology Design* in which she provides us with different situations where different users of the mobile phone device adopted and localized a hard-to-use technology to fit their everyday situations. Based on the case stories, she re-conceptualizes localization by advancing a concept known as user localization. This concept stresses that localization is both a situated action and a constructed meaning. This means that contextual factors such as weather, local politics, local histories, local logics, local needs, and economic situations shape localization processes. In addition, she proposes that design philosophy and principle should attend to "both action and meaning through a cyclical design process." This, she argues, will help "the cross-cultural design community to create a technology both usable and meaningful to local users." The core tenet of user localization is "to link design to use" (Sun, 2012, p. 42).

This, to me, is a worthy addition to localization scholarship because it shifts attention from the idea that localization focuses on translation of texts to local languages or merely changing graphical interface for users (Esselink, 2000; Fukuoka, Kojima, & Spyridakis, 1999; Hoft, 1995; Taylor, 1992; Thayer & Kolko, 2004) to a careful attention to the sociocultural and atmospheric contexts surrounding use situations. As she identifies, localization specialists only focus on delivery aspects, that is, "specialists only pay attention to such issues as colors, page layouts, or how dialogue boxes should be resized for a certain language" (Sun, 2006, p. 460) and this leads to a neglect of the broader sociocultural contexts where "information products are situated, and where products are designed, produced, distributed and consumed" (Sun, 2004, p. 2).

Sun's user localization, more or less, is a reinvention of Johnson's idea of user-centered design although both scholars differ in their approaches—Sun is more concerned about design for cross-cultural users from the point of localization, while Johnson is more interested in advancing user theory from a rhetorical perspective; users control design through strategies in Sun's book, while "users do not actually control the design in Johnson's case" (Spinuzzi, 2003, p. 8)—both scholars emphasize the importance of paying attention to user knowledge, user strategies, the complex nature of use, and the situatedness of technology. My book seeks to contribute to conversations about localization by examining the various ways Ghana localized the biometric verification registration (BVR) and BVD adopted to enhance the country's electoral system for the 2012 presidential and parliamentary elections and subsequent elections.

My study of Ghana's use of the biometric and published literature on biometric use in elections indicate that biometric technology designers and users must make localization a top priority if these technologies will satisfy user needs. These stories which expose the inabilities of biometric technology provide enough grounds to study user situations surrounding the use of this technology, particularly since it is being adopted by many electoral systems. I hold that users' interaction with the biometric will unravel decisions and actions that "undermine the theoretical potentials or promises of 'anti-corruption' technologies" (Hobbis & Hobbis, 2017, p. 115) such as the BVD used by Ghana. Hobbis and Hobbis (2017) state, and I agree that this kind of study "allows us to move beyond technologically deterministic understandings that assume the (effective) deployment of a given technology will lead to a particular outcome (improved voter integrity and possibly a more stable electoral and political system)." In the coming pages of this book, I will reveal how the discourse surrounding the biometric adopted by Ghana maintains, articulates, and rearticulates a technologically deterministic understanding of the technology. I refer to this deterministic portrayal of the biometric as *the biometric ideology*: the belief that the technology is neutral, objective, accurate, and truthful. I use "ideology" with a caution. I am not saying that the biometric used in Ghana did not ensure some success or that we cannot rely on it—in fact, most of the election officials I interviewed emphatically applauded the efficacy of the biometric device used; I only mean that the biometric was communicated in a way that focused on deterministic discourses which only communicated the instrumental features of the technology. I intend to portray the consequences that come with solely designing communications that seek to focus on instrumental features of a specific technology.

Theorizing Localization: Linguistic, Subversive, and User-Heuristic Localization

I reveal that successful use of a technology does not lie in the use of persuasive language in describing a technology; rather, success is judged by UX and the various heuristic approaches implemented by users to make technology fit local context. Persuasive use of language, to an extent, contributes to success in sales and not user satisfaction. My belief is grounded in the idea that technology users are not passive or a bunch of dummies; rather they are the heroes who "shape, redesign, and localize an available technology to fit into their local contexts" (Sun, 2012, p. xiv). I call this effort made by users to become heroic users of technology *participatory user localization,* and this can be expressed in three different ways: *user-heuristic localization, linguistic localization, and subversive localization.* Participatory user localization borrows from Sun's (2012) idea of "user localization," that is, efforts made by users to incorporate technology into their daily lives, and Agboka's (2013), "participatory localization," that is, "user-in-community involvement and participation in the design phase" (p. 42), as a starting point, but I consider my project, although related, considerably different in that whereas user localization and participatory localization emphasize daily lives, and user participatory localization focuses on how users localize technology by using either heuristic approaches to resolve technological problems (user-heuristic), subverting, and redesigning documents surrounding technologies adopted (subversive localization) or drawing on deterministic discourses to persuade users of the powerfulness of the biometric adopted (linguistic localization).

Linguistic localization, as I advance in this book, helps me to highlight connections and similarities with the *rhetoric of technology*. I establish this relationship because technical communication is rhetorical (Rutter, 1991; Spinuzzi, 2003), and also because rhetoric has been used to set limits, prescribe, and define what technology is; it has been used to upend our notions of technology (Haas, 2012, p. 287). Thus, rhetoric of technology, defined as "…the rhetorical productions that surround a material technology" (Bazerman, 1998, p. 381), such as the biometric or "the rhetoric that accompanies technology and makes it possible—the rhetoric that makes technology fit into the world and makes the world fit with technology" (p. 385), helps me to highlight the instrumental discourses that shaped the description of the biometric. According to Bazerman, technologies are always already rhetorical because they are developed to meet the needs of humans and they are always "part of human needs, desires, values, and evaluation, articulated in language and at the very heart of rhetoric" (p. 383). More so, technologies must go through several public for approval before they become a material reality. For instance, in Ghana's case, the biometric had to be approved by the parliament of Ghana, Inter-Party Advisory Committee (IPAC), and civil society organizations. To this end, the technology must argue for value in regard to business, law, government, the public, and consumers.

Before Bazerman, Miller (1994) argued that language is used persuasively to maintain things technical. *Linguistic localization* captures the extent to which things technical become part of human consciousness because of the perceived notion that they are neutral, efficient, objective, accurate, and provide accurate results. Katz (1992) refers to this trust in things technical as the *ethic of expediency*, that is, a phenomenon where "science and technology become the basis of a powerful ethical argument for carrying out any program" (p. 264). When this happens, technology becomes both a means to an end and an end itself. As I will demonstrate in my fifth chapter, linguistic localization works to meet the cultural expectation of users by presenting instrumental features of a technology but not addressing complex activities such as how weather conditions affect technology use, unethical polling officers who work to massage figures to favor their political cronies as was the case in Ghana, and poor laws that endanger electoral processes, or the perception that going through biometric verification will render males impotent and destroy pregnancy. I reveal that over-reliance on deterministic discourses which favor instrumentality could result in poor UX of adopted technologies.

When users realize, for instance, that genres that are imposed on them do not capture "local contingency," (Spinuzzi, 2003, p. 3) they pick it up, reinvent the genre, and use it to their needs. I call this effort *subversive localization*. As I will discuss in Chapter 6, when the EC adopted the biometric technology, officials of the commission realized that the manual accompanying the biometric technology did not say anything about how the biometric will be used to register voters and illustrations did not help users to understand how to use the biometric in the electoral process, so the commission had to redesign the user manual to meet local needs, that is, the electoral process of Ghana. I indicate that the users in Ghana are able to subvert and redesign a new user manual because they possess "practical wisdom," that is, "a true and reasoned state of capacity to act…" (McKeon, 2009, p. 1026).

The final type of localization I discuss is *user-heuristic localization*. At the center of user-heuristic experience, localization is the idea that user knowledge is not regulated by formal rules and procedures learned in the formal education system or through any formal means; instead, users develop their expertise and form of problem-solving in response to unpredictable real-life situations that arise in regard to technology use. For instance, when the biometric technology started failing on Election Day, the EC had to come up with quick fixes to solve the issues. How did the EC and voters know that when an individual is rejected by the technology, the person could use Coca-Cola to clean his/her hands? Interestingly, a report which captures the success of the heuristic approaches used by the EC to solve breakdowns labeled the approaches as "unscientific." This label, for me, is an attempt to devalue user knowledge that emerged to partially solve biometric breakdowns. Regardless of the categorization, I consider this move to use Coca-Cola, OMO detergent, and other simple means to resolve issues and get voters to vote a problem-solving activity. These processes I discuss reveal the various ways users were able to make sense of the situation and creatively customized and localized a technology to fit into the electoral system, contributing to their identity-building process. Thus, localization should not solely be about how designers change technological features to meet local situations; users also influence design decisions at micro-levels.

I have stated earlier that I am more concerned about the users and their creative activities which ensured integration of the technology. Three reasons account for this shift from designers to users: First, creative approaches used by users to localize biometric technology to fit their local electoral systems is a "relatively unexplored land" (Johnson, 1998); second, many researchers have focused on designers and developers of biometric technology; and third, the difficulty EC officials had in describing or talking about the designers of the biometric technology which was adopted and used. This experience reveals that in most cases, even though users don't know who designed or developed the technology, they nevertheless manage to use the technologies effectively. The electoral officials I interviewed could not state the company which designed the biometric the country adopted; neither were verification officers able to reveal the designers of the user manuals they used during the elections. They expressed the desire to be part of the design process, but they also realized that they could not be part of it because these are decisions made by the top hierarchy of the Commission.

What is the implication of this gap for industries which seek to localize their technologies and users who seek to be part of the design process? What is the implication for technical communication scholars who seek to make users the center of design processes? This is a wicked question designers of biometric technologies and technical communicators must work to respond. The global interconnectedness of the twenty-first century makes thinking about users quintessential because the success of a technology will depend on how people from different regions of the world understand, adopt, and use a product. More importantly, interest in biometric technology will keep soaring when users remain interested in their products "not only because it looks good, but because it keeps the users engaged in continuous exploration, newer applications, a deeper level of personal satisfaction, dynamic feedback, and a sustained dialog among users of the system" (Roy, 2013, p. 112).

Since it is a fact that users may never meet designers and vice versa, I will not attempt to bridge the gap because it will be of little relevance, at least for this research, to discuss designers, but I do not disregard their influence in shaping technology for users and I do not disregard design issues entirely. Sun's scholarship (see 2006, 2009b, and 2012), for instance, indicates that it is necessary for designers of technology to connect design to use because there is almost always a complex interaction between the domain of design and the domain of use. She explains that when there is a separation between design and use, the technology becomes detached from its context. As soon as this detachment is established, a gap is created between these two sources. The consequence is a breakdown of the technology being used. This argument holds true to an extent considering that in Ghana's situation, the country experienced numerous breakdowns when it first used the biometric in 2012. This could be attributed to the lack of communication between users and designers of the biometric. In subsequent elections, however, Ghana registered a number of successes although users did not co-design the biometric with designers: The EC recorded fewer breakdowns, and officials adopted measures to handle voters who were rejected by the device. They also adopted strategies to deal with the overheating of biometric technologies. The picture painted here is that the designer is of little relevance. It is rather more fulfilling to focus on the users, their experiences, their activities, their needs, and aspirations to find out how they were able to overcome technological challenges. In this way, users will be cast as experts who initiate strategies to save technology from further jeopardizing an already deteriorating electoral system.

The Wicked Problem for Technical Communicators in the Era of Biometric Elections: Testing Our Theories in a Global Village

While my prime focus is on localization processes, my study is situated in a broader concern: election technology in a non-Western context. I am aware that the subject of elections or biometric use in elections as discussed in this book is an uncharted path in the field of technical communication. Some may even think that I am overstepping the boundaries of technical communication scholarship. It may be true, but I am equally aware that technical communicators are versatile enough to trod uncharted paths. For me, this ability to discuss a variety of issues is our strength, and Johnson (1998) is right in saying that "technical communication already functions interdisciplinarily—this is one of its strongest assets..." (p. 14) and a weakness (Rude, 2009, p. 177). My project aims to engage in arguments that seek to advance civic engagement, localization, social justice, and technology studies in our field. Over the years, concerns for user-centered design, human rights, and human dignity are increasingly being championed (Sapp, Savage, & Mattson, 2013;

Walton, 2016) in our field. So we have calls for civic participation or public intellectualism skyrocketed (Bowdon, 2004; Dubinsky, 2004; Simmons & Grabill, 2007; Walton, 2013a). We have discussed human rights, social justice, local–global tensions, ethics, and good citizenship. By participating in these broad conversations, we have demonstrated that we are part of a larger society and that we need to contribute to societal growth. In fact, we have become that "participatory citizen" Johnson (1998) imagined some decades ago. Therefore, we cannot pretend electoral issues have no significant place in our scholarship, particularly when these electoral processes are managed by institutions that are headed by individuals and each individual has "his or her own tactical response to institutional power" (Kimball, 2017, p. 3).

Voting processes involve the use of several technologies (defined broadly to include paper ballots, voter registration, voter education materials, pink sheets,[2] Internet technologies, scanners, ballot printing materials, user manuals, and other technical documents used during elections) which use language in one way or the other. I believe that a study of biometric use in electoral processes can enable scholars in our field to expand our research beyond organizations to include different contexts, activities, and spaces (Kimball, 2006; Rose, 2016; Seigel, 2013). Such a study can also enable scholars to appreciate the role that technical documents play in sustaining assumptions about a specific technology. As ardent communicators of technology to users, we will be able to understand how cultural, social, rhetorical, and local factors affect the communication and adoption of a technology. More importantly, such a study helps to examine the various ways elections have been subjected to the logic of technology; it provides an exigence to articulate the historical, political, social, rhetorical, and cultural entanglements surrounding the communication of a technology in a non-Western context. More so, such a study establishes a fluid relationship among technical communication, technology, elections, and democracy; it brings elections and democracy into the domain of technical communication. And we cannot disregard the fact that a study based on localization helps the field to advance the idea that users have knowledge of use; hence, they must be seen as experts in their own rights.

I would also want to point out that although we do not discuss elections per se in our field, some scholars have given a hint to its relevance to our scholarship. Agboka (2013) implicitly hints at the possibility of discussing electoral issues in technical communication when he asked scholars to expand research to dis/enfranchised cultural sites. Though Agboka uses "disenfranchise" to refer to intercultural and international contexts, I interpret it here to mean depriving someone of the right to cast a ballot. That word, fortuitously, points to electoral-related concerns. More recently, a special issue edited by Dorpenyo and Agboka (2018) on "election technology, technical communication and civic engagement," received overwhelming support from TC scholars. The contributors to the special issue indicate that technical communicators can discuss election-related matters.

Perhaps biometric technology use in elections in resource-mismanaged areas[3] such as Ghana provides enough grounds for us to expand technical communication practice and research. The adoption of the biometric technology in Ghana and the reasons for the adoption point to the fact that technology use is not specific to the Western world; it is, I would say, a universal case. We could say that Westerners and non-Westerners have similar ideologies or assumptions about technology: It is a quick fix to our problems, and it provides efficient results. Therefore, we must study how cultures other than Western cultures relate to technology. This will broaden the purview of technology studies, research, and arguments made about technology. Just like Bernadette Longo (2000), I argue that we should study how the technology works in non-Western or international contexts. This call should not be left to a group of scholars. This is an avenue for technical communicators to broaden their research horizon because the world has become a globalized society; technology is used by international and cross-cultural audiences.

While technical communicators are user advocates, the important question is: How can we advocate that designed technologies should "meet the cultural [and I add, rhetorical] expectations of local users, support their complex activities in concrete contexts, and empower agency and mediate their identities in" (Sun, 2012, p. 4) non-Western societies? And how can we make an active case for user knowledge to be included in technology design processes? We need to investigate these issues by conducting and reporting case studies that consider the various ways users outside Western societies relate to and use technology. Huatong Sun has taken a bold step in this direction with her study of mobile technology use in both American and Chinese cultures. In this study, she identifies that technology design processes create "a disconnect between action and meaning" of local users (Sun, 2012, p. 8). That is, little is done to incorporate the meanings, behaviors, practices, and state of consciousness of users into technology design. As a result of this neglect, she proposes that design philosophy and principle should attend to "both action and meaning through a cyclical design process." This, she believes, will help "the cross-cultural design community to create a technology both usable and meaningful to local users." There is a need to push for a user localization, that is, a study of the various ways that users localize technology to fit everyday lifestyles (Sun, 2006, p. 459), as this provides an insight into local forms of life and how we can adapt or design technologies to meet user needs.

Similarly, Agboka (2013) advocates a "participatory localization" in the design of documents and technologies in international or cross-cultural contexts. By "participatory localization," the author refers to the extent to which users will be involved in design phases. The user who participates in localization processes is not an isolated person, but he/she is involved as a member of a community (p. 42). He identifies that though technical communication scholarship is replete with ideas and notions of localization, little is done to account for user needs in resource-mismanaged sites. Localization "often starts from the developer's site and trickles down to the users' site" (p. 30). Thus we can conjecture that designs and technologies that are sent to unenfranchised sites reinforce the system-centered design model discussed by Johnson in *User Centered Design*.[4] Again we can contend that designs carry the stigmata of the developers. User advocacy and localization in international context is a ripe domain for technical communicators to study technology use.

It is interesting, however, to note that research has neglected political contexts of technology use in non-Western societies. Specifically, we have not bothered to study election technologies such as the biometric technology. For instance, Rosario Durao (2013) who makes an attempt to map out the various thematic strands that run through international technical communication/professional research, identifies five interesting dimensions—environment, human, communication, technology, and artifacts. Under the technology dimension, Durao reveals that discussions on technology are connected to "equipment, professional contexts, educational contexts, entertainment contexts… viewpoints" (Durao, pp. 18–19). Nothing is said about election-related contexts of technology. This silence on this aspect of the political is probably due to the fact that technical communication as a field has shied away from discussing issues relating to elections. This is ironic since technical communication scholars cherish conversations about civic engagement. To the best of my knowledge, election-related concerns can be considered civic issues. In fact, the UN's declaration on human rights makes voting a prime civic responsibility of every individual in the society. This neglect means we have not fully heeded to Blyler's (1998) call that the field ought to take a political turn in its research practices. My research to a larger extent examines the concatenation between politics and technology in an international context. I have embraced Blyler's and Agboka's calls. My work moves away from the narrow focus of research on workplace contexts "toward serious engagement with under-examined issues such as" (Ding & Savage, 2013, p. 1) localization, election technology, user strategies, biometric failures, and human-technology relationship in a non-Western society.

We may not be so much interested in election-related issues, but we have some few people to look up to: Justin Whitney (2013) demonstrates how technical communicators can contribute to electoral issues when he critically studied The 2010 Citizens Clean Elections Voter Education Guide designed with the purpose of providing Arizona voting public with the needed information about state elections and found out how discourses used in this document construct immigration or immigrants as "illegal" and a "social problem" which requires a "fight" or a "crack down" (p. 438). He admonished that we have a role in helping electorates to understand how personal political gains shape how information is communicated and how technical communications construct "a perceived identity of certain groups of people and influences the reactions that other groups have to them" (p. 451). Although the domain of elections is traditionally carved for scholars in the field of political science, Whitney demonstrates that we can "apply simple rhetorical concepts" (Bowdon, 2004, p. 333) to solve electoral problems. Dorpenyo and Agboka (2018) special issue on technical communication and election technologies also reveal how technical communication theories can be helpful in discussing election-related concerns.

José (2017) also brings electoral issues to technical communication when she conceived of the "Brexit vote as a technical communication failure" and emphasized that technical communication scholars wield enough power to interpret "current political events…" Whitney Quesenbery and Dana Chisnell, co-directors of the Center for Civic Design, are using their skills and expertise in UX to design electoral ballots to help electoral systems throughout the US. Chisnell, for instance, "brings deep experience in civic design, starting with research at NIST into the use of language in instructions on ballots (with Ginny Redish), and work on standards and testing for poll worker documentation for the Voluntary Voting System Guidelines (VVSG)." Democracy and [by extension election] is a "design problem" ("We Believe that Democracy is a Design,") and technical communicators are designers, so we need not be political scientists in order to contribute to conversations about elections, especially when technology plays a central role in an electoral system.

Apart from the silence on the political, in terms of elections, there is also neglect on discussing the political nature of technologies adopted in non-Western contexts. By this I mean, little is done to discuss how technologies adopted or exported to non-Western contexts are complicit in racial discrimination or how such technologies are not neutral and are shaped by the ideologies that are embedded into their design. While I spend a lot of time discussing why the biometric breakdowns in Ghana help to theorize and re-think localization process, I argue in Chapter 8 that there is a social justice component to localization and that if we will make gains in localization theories, we must discuss localization vis-a-vis social justice. In Chapter 8, I will reveal how localization and social justice are intrinsically linked. For instance, interviewees and technical reports which captured election proceedings indicated that the biometric technology used struggled to read the fingerprints of specific groups of people. Yankson hinted that "all those who had problems... if it's not the aged those who are very old and the person is young, then is per the work that he/she does or what he or she uses the fingers to do except for those who were very old." Besides these claims by my interviewees, some technical documents captured the rejection of voters by the biometric technology. A report by the Commission on Human Rights and Administrative Justice (CHRAJ)[5] indicates that "most of those affected by the problems of verification were elderly women, some young women who had dyed their hands and fingers with a local herb called 'lenle' were also affected." These examples, which will be elaborated on in Chapter 8, indicates the need to interrogate the social justice component of biometric use. It is not enough to point out localization strategies, and it is equally rewarding to signify instances of injustice embedded in technology use.

Technical Communication in the Age of Biometric Technology

Technical communication scholars have been fascinated by the study of technology at different levels. Therefore, this study is an extension of literature in the intersection among technical communication, cultural studies of technology, and technology studies (Johnson, 1998; Kitalong, 2000; Longo, 2000; Paradis, 2004; Seigel, 2013; Sun, 2006, 2009a, b, 2012; Walton, 2013b). The question thus lies in how technical communicators can study a technology such as the biometric. If a study of the biometric is possible, then where can scholars begin their search?

One of the ways to understand a particular technology and its social significance is to study the technical documents that order, shape, and produce technological knowledge. I derive my understanding of Dobrin's definition of technical communication as "accommodation of technology to users" (Dobrin, 2004) from the standpoint that technical documents are very powerful and their powerfulness enable us to help users to make meaning of, interpret, understand, and use technological systems. Johnson (1998), Paradis (2004), and Seigel (2013) speak powerfully about how instructional aspects of technology are within the domain of technical communicators, and it is by looking at this domain that we can add to scholarship and contribute to large social conversations.

One cannot dispute the relationship between technical communication (broadly defined to include technical/professional writing) and technology (systems, objects, knowledge, and practices). Longo (2000), as an example, indicates the extent to which technical writing gained ascendancy as the "currency" for scientific and technical knowledge; it "controls how technical knowledge is made" (p. x). Through documents, such as reports, technical writers are able to order knowledge, shape information, and make arguments about what is to be valued. In "What can History Teach us About Technical Communication," Longo and Fountain (2013) make an explicit statement about our field:

> technical communication operates as a system of ordering that shapes and produces technological, scientific, and often business knowledge through the creation and use of texts; the processes of ordering often take the form of organizational culture or institutional history, which continue to influence the contemporary workplace environment; through the influences of the past and the present, technical communication practices and documents play a socializing role; they encourage or value certain practices

while discouraging or complicating others; and traces of history and the socializing function are witnessed in the 'actual documents of a workplace' (p. 174). The most important thing to recognize as technical communicators is that "we need to know not only the history of a document, the people who created it, and its current usage, but also the role we play in the larger social practices.... (p. 182)

I will reveal throughout this book that through technical writing, that is, writing which "accommodates technology to users" (Dobrin, 2004), the EC and stakeholders created rhetorical frames and appeals that helped persuade Ghanaians to accept the biometric technology as the surest way of producing authentic electoral results. As I will advance later in my book, my concept of linguistic localization will heavily focus on the technical documents designed by Ghanaians to persuade Ghanaians. I will pay attention to how language used in such documents, sought to control knowledge about the biometric device which was adopted; but I will also reveal the social role of such documents. To this end, documents such as the biometric verification user manual, parliamentary hansard,[6] suite of posters, and an online video used by the EC become very relevant to this conversation. Stakeholders dwelt on extra-rhetorical tools such as malpractices that have characterized electoral culture in Ghana to make a powerful case for the adoption to be possible. Technical writing in this regard becomes the medium or instrument through which institutions, societies, organizations, and individuals create exigence for a technology to be adopted. It also becomes a knowledge-producing practice.

We also must capture lived or users' live experiences by conducting research on rigorous qualitative interviews to enable users to tell their stories. I combine textual analysis and qualitative coding of interview data coupled with a decolonial methodology to theorize localization and reveal the social justice implications of the biometric technology adopted by Ghana. Overall, biometric technology has come to assume an unprecedented position in the electoral process of Ghana and several other countries in Africa, and it also shapes our understanding about how non-Western societies make sense of, integrate, and communicate about technologies that are adopted. It is an avenue to investigate how knowledge about technologies adopted are legitimized and produced and the role technical documentation plays in the communication processes, how the biometric use in Ghana will advance localization theory and practice, how local knowledge making about technology in decolonial contexts are masked in colonial rhetorics, what role technical documentation plays in unmasking assumptions about technology, and how rhetoric interacts with cultural, ideological, historical, and political forces to shape the representation or communication of technologies adopted.

Overview of Chapters

Chapter 1 serves as the introduction to the manuscript. In this chapter, I lay a foundation for the discussion of the three localization processes I advance in the book: linguistic localization, subversive localization, and user-heuristic localization. Chapters 5, 6, and 7 fully develop these concepts, respectively. Ultimately, I argue that these three processes help to capture users as heroes who shape, redesign, and localize an available technology for their need. I am also able to discuss the subject of the book: user localization strategies; discuss the aims of the book, the audience, significance, and the major questions that drive the project.

Chapter 2 contextualizes the project by discussing two main issues: the political history of Ghana as well as some issues that drive political circumstances in Africa and the biometric technology as a global technology that ensures authenticity. After I construct a larger context for the country's contemporary political situation, I review the literature on biometric technology. I discuss why it has become a global technology and how its affordances can (perceivably) provide some kind of hope for those countries that want to enhance their electoral integrity. The political history of Ghana and the brief description of the biometric will help us all to appreciate the global logics which pushed the EC and other stakeholders to adopt this specific technology to solve a local problem. These overviews help me to elucidate the question: How did Ghanaians or the EC decide to adopt a biometric verification system?

After thoroughly grounding my localization processes and the context of the research, Chapter 3 turns to and establishes the methodological framework for the study of localization and technology use in non-Western, colonized contexts. International technical communication scholars have identified the complexity in studying international, intercultural, or cross-cultural context and have advocated the use of methodologies that can capture the complex nature of the context under study. In this regard, I propose the use of decolonial methodologies and its accompanying methods of grounded theory and rhetorical–cultural analysis. I believe decolonial methodology is the appropriate technology and, in this chapter, I lay out the various theoretical principles guiding this methodology. Decolonial methodology seeks to recover the lost identities of colonized people by championing self-determination, empowerment, decolonization, and social justice. I see decolonial methodology as the appropriate approach to conduct research that seeks to advance user localization strategies for two reasons: (1) users' ways of knowing have been marginalized and colonized by discourses that favor designers. This means we need to recover the lost voices of users; and (2) the context of study, Ghana, is a colonized context, so it is necessary to use an approach which will help me to demystify colonialism. Because decolonial is a flexible methodology, I am able to combine it with grounded theory method. Grounded theory enables me to respect and pay attention to my research data and it gives me the ability to construct theory from ground up. Building knowledge from below is a core principle of a decolonial project.

Chapter 4 introduces four specific individuals I interviewed and their experiences with the biometric technology to contextualize theoretical developments. Reading all four cases in entirety provides a more complete vision of the user localization strategies I advance in this book. The individuals are carefully selected to reflect the different structures of the EC and the different levels of authority and how each level encountered the biometric technology.

The first of the three localization processes, linguistic localization, is discussed in Chapter 5. I analyze an ecology of genres that accompanied the biometric device adopted by the EC: a user manual, a Parliamentary Hansard, suite of posters, and an online video designed. My goal is to identify the extent to which the documents are maintained and constituted the logic of the biometric technology. I use the term the "logic" of biometric to refer to the ideology that biometric provides an effective fix to electoral problems. I argue that in most cases, technical genres are used as the medium to represent the instrumental features of technologies. What is fascinating is the fact that the documents which accompanied the device adopted was designed by Ghanaians for Ghanaians.

Chapter 5 communicates how the biometric should work, and Chapter 6 discusses how the biometric actually worked and how users devised various tactics to resolve biometric failures. When the biometric technology started failing on Election Day, the EC had to come up with "unscientific" or vernacular methods to prevent the technology from jeopardizing the electoral process. This chapter focuses on the heuristic approaches used to resolve biometric breakdowns to discuss what I term *user-heuristic experience localization*. At the center of user-heuristic experience, localization is the idea that user localization is not regulated by formal rules and procedures learned in the formal education system or through any formal means; instead, users develop their expertise and form of problem-solving in response to unpredictable real-life situations that arise in regard to technology use. Ghana's use of the biometric technology indicates that such user strategies as stepping back, learning and experience are central to localization approaches. Users do not get to acquire these traits easily, through educational lectures or reading, but laboriously through past mistakes/encounters. At the level of user-heuristic experience, user moves away from linguistic prestidigitations to practical forms of knowledge making; they think about broad contextual use situations of a technology than a mere use of flamboyant language to describe the functional attributes of a technology (linguistic localization).

Chapter 7 also presents another use of tactics by users by extending Spinuzzi's concept of subversion. I use subversive localization as a way of discussing the instant when a user picks "up available tools, adapts them in idiosyncratic ways, and makes do" (Spinuzzi, 2003, p. 2). As I will discuss in this chapter, when the EC adopted the biometric technology, officials of the commission realized that the manual accompanying the biometric technology did not say anything about how the biometric will be used to register voters and illustrations did not help users to understand how to use the biometric in the electoral process so the commission had to redesign the user manual to meet local needs, that is, the electoral process of Ghana. I indicate that the users in Ghana are able to subvert and redesign a new user manual because they possess "practical wisdom," that is, a true and reasoned state of capacity to act.

Chapter 8 takes a step back to reflect on how the biometric failures provide an avenue to discuss social justice issues. This chapter argues that although the biometric technology is touted to be value-neutral, objective, and accurate, it is inherently discriminatory. As I indicated in previous chapters, reports show that the biometric rejected those individuals who are engaged in "slash-and-burn agriculture." Therefore, the mass subjection of elections to the logic of the biometric technology in resource-mismanaged contexts is welcoming, but its use raises social justice and localization concerns. I will indicate the relevance of social justice to decolonial and localization projects.

The concluding chapter presents broader implications of the empirical study of Ghana's use of biometric in elections. I further develop the framework of participatory user localization with a discussion of interview findings and analysis of genres accompanying the biometric technology. I also discuss future directions for the research and practice of user localization strategies in a globalized era. Focusing on a decolonial approach, the chapter analyzes what the design community could learn from the user localization strategies I advance to design for, involve, nurture, pay attention to, and sustain technology designs which are sensitive to user knowledge and local logics.

Notes

1. All interviewee names used are pseudonyms.
2. In Ghana's political context, the term refers to "the form on which the Statement of Poll and Declaration of Result for the office of President and Parliament [is] recorded" (Baneseh, 2015, p. xi).
3. I choose resource-mismanaged context over other terms as "resource constrained contexts" for the reason that most on these non-Western countries are not constrained in any way. Rather, resources have been mismanaged by myopic, selfish leaders, and policy makers. Ghana, for instance, has gold, cocoa, timber, oil, manganese, bauxite, water bodies, and human capital in abundance. We wallow in poverty because we have not prioritized properly.
4. For more on system-centered design, read Robert Johnson (1998).
5. CHRAJ is an independent organization established under the 1992 Constitution of Ghana. Its core function is to ensure that ensure that individual human rights and freedoms are protected and promoted in the country.
6. Refers to "the official nearly verbatim report of proceedings of [Parliament]. It is a repository and reflection of the legislative activities of Parliament" (http://www.parliament.gh/publications/).

References

Agboka, G. Y. (2013). Participatory localization: A social justice approach to navigating unenfranchised/disenfranchised cultural sites. *Technical Communication Quarterly*, 22(1), 28–49. https://doi.org/10.1080/10572252.2013.730966.

Baneseh, M. A. (2015). *Pink sheet: The story of Ghana's presidential election as told in the Daily Graphic*. Accra: G-Pak Limited.

Barton, B. F., & Barton, M. S. (2004). Ideology and the map: Toward a postmodern visual design practice. In J. S. Johnson-Eilola & S. A. Selber (Eds.), *Central works in technical communication* (pp. 232–252). New York: Oxford University Press.

Bazerman, C. (1998). The production of technology and the production of human meaning. *Journal of Business and Technical Communication, 12*, 381–387.

Blyler, N. (1998). Taking a political turn: The critical perspective and research in professional communication. *Technical Communication Quarterly, 7*(1), 33–52.

Bowdon, M. (2004). Technical communication and the role of the public intellectual: A community HIV-prevention case study. *Technical Communication Quarterly, 13*(3), 325–340. https://doi.org/10.1207/s15427625tcq1303_6.

Brady, A. (2004). Rhetorical research: Toward a user-centered approach. *Rhetoric Review, 23*(1), 57–74.

Colman, J. (2015). *Wicked ambiguity and user experience*. Retrieved from http://www.jonathoncolman.org/2015/05/21/wicked-ambiguity/#video.

Ding, H., & Savage, G. (2013). Guest editors' introduction: New directions in intercultural professional communication. *Technical Communication Quarterly, 22*(1), 1–9.

Dobrin, D. (2004). What's technical about technical writing. In J. S. Johnson-Eilola & S. A. Selber (Eds.), *Central works in technical communication* (pp. 107–123). New York: Oxford University Press.

Dorpenyo, I., & Agboka, G. (2018). Technical communication and election technologies. *Technical Communication, 65*(4), 349–352.

Douglas, T. (2017). To design better tech, understand context. *TEDGlobal 2017*. Retrieved from https://www.ted.com/talks/tania_douglas_to_design_better_tech_understand_context/transcript?language=en#t-503002.

Dourish, P. (2003). The appropriation of interactive technologies: Some lessons from placeless documents. *Computer Supported Cooperation Work, 12*, 465–490.

Dubinsky, J. M. (2004). Guest editor's introduction. *Technical Communication Quarterly, 13*(3), 245–249.

Durao, R. (2013). International professional communication: An overview. *Connexions: International Professional Communication Journal, 1*(1), 1–24.

Esselink, B. (2000). *A practical guide to localization* (Vol. 4). Amsterdam: John Benjamins Publishing Company.

Fukuoka, W., Kojima, Y., & Spyridakis, J. H. (1999). Illustrations in user manuals: Preference and effectiveness with Japanese and American readers. *Technical Communication, 46*(2), 167–176.

Gonzales, L. (2018). *Sites of translation: What multilinguals can teach us about digital writing and rhetoric*. Ann Arbor: University of Michigan Press.

Gonzales, L., & Zantjer, R. (2015). Translation as a user-localization practice. *Technical Communication, 62*(4), 271–284.

Haas, A. M. (2012). Race, rhetoric, and technology: A case study of decolonial technical communication theory, methodology, and pedagogy. *Journal of Business and Technical Communication, 26*(3), 277–310.

Hobbis, S. K., & Hobbis, G. (2017). *Voter integrity, trust and the promise of digital technologies: Biometric voter registration in Solomon Islands.* Paper presented at the Anthropological Forum.

Hoft, N. L. (1995). *International technical communication: How to export information about high technology.* New York, NY: Wiley.

Johnson, R. (1998). *User centered technology: A rhetorical theory for computers and other mundane artifacts.* Albany: State University of New York.

Jones, N. N. (2016). Narrative inquiry in human-centered design: Examining silence and voice to promote social justice in design scenarios. *Journal of Technical Writing and Communication, 46*(4), 471–492. https://doi.org/10.1177/0047281616653489.

José, L. (2017). *User-centered design and normative practices: The Brexit vote as a technical communication failure.* Paper presented at the Association of Teachers of Technical Writing—Twentieth Annual Conference, Portland, OR.

Katz, S. B. (1992). The ethic of expediency: Classical rhetoric, technology, and the Holocaust. *College English, 54*(3), 255–275.

Kimball, M. A. (2006). Cars, culture, and tactical technical communication. *Technical Communication Quarterly, 15*(1), 67–86.

Kimball, M. A. (2017). Tactical technical communication. *Technical Communication Quarterly, 26*(1), 1–7. https://doi.org/10.1080/10572252.2017.1259428.

Kitalong, K. S. (2000). "You will": Technology, magic, and the cultural contexts of technical communication. *Journal of Business and Technical Communication, 14*(3), 289–314.

Longo, B. (2000). *Spurious coin: A history of science, management, and technical writing.* Albany: State University of New York Press.

Longo, B., & Fountain, K. (2013). What can history teach us about technical communication? In J. Johnson-Eilola & S. A. Selber (Eds.), *Solving problems in technical communication* (pp. 165–186). Chicago: The University of Chicago Press.

McKeon, R. (2009). *The basic works of Aristotle.* New York: Random House LLC.

Miller, C. R. (1994). Opportunity, opportunism, and progress: Kairos in the rhetoric of technology. *Argumentation, 8*(1), 81–96.

Paradis, J. (2004). Text and action: The operator's manual in context and in court. In J. S. Johnson-Eilola & S. A. Selber (Eds.), *Central works in technical communication* (pp. 365–380). New York: Oxford University Press.

Report of the Commonwealth Observer Group: Ghana Presidential and Parliamentary Elections. (2013). Retrieved from London. http://thecommonwealth.org/sites/default/files/inline/GhanaElections-FinalReport2012.pdf.

Rose, E. J. (2016). Design as advocacy: Using a human-centered approach to investigate the needs of vulnerable populations. *Journal of Technical Writing and Communication, 46*(4), 427–445. https://doi.org/10.1177/0047281616653494.

Roy, D. (2013). Toward experience design: The changing face of technical communication. *Connexions: International Professional Communication Journal, 1*(1), 111–118.

Rude, C. D. (2009). Mapping the research questions in technical communication. *Journal of Business and Technical Communication, 23*(2), 174–215.

Rutter, R. (1991). History, rhetoric, and humanism: Toward a more comprehensive definition of technical communication. *Journal of Technical Writing and Communication, 21*(2), 133–153.

Salvo, M. J. (2001). Ethics of engagement: User-centered design and rhetorical methodology. *Technical Communication Quarterly, 10*(3), 273–290.

Sapp, D. A., Savage, G., & Mattson, K. (2013). After the International Bill of Human Rights (IBHR): Introduction to special issue on human rights and professional communication. *Rhetoric, Professional Communication, and Globalization, 14*(1), 1–12.

Scott, B. (2003). *Risky rhetoric: AIDS and the cultural practices of HIV testing*. Carbondale: Southern Illinois University Press.

Seigel, M. (2013). *The rhetoric of pregnancy*. Chicago: University of Chicago Press.

Simmons, W. M., & Grabill, J. T. (2007). Toward a civic rhetoric for technologically and scientifically complex places: Invention, performance, and participation. *College Composition and Communication, 58*(3), 419–448.

Spinuzzi, C. (2003). *Tracing genres through organizations: A sociocultural approach to information design* (Vol. 1). Cambridge, MA: MIT Press.

Sun, H. (2004). *Expanding the scope of localization: A cultural usability perspective on mobile text messaging use in American and Chinese contexts*. Troy, NY: Rensselaer Polytechnic Institute.

Sun, H. (2006). The triumph of users: Achieving cultural usability goals with user localization. *Technical Communication Quarterly, 15*(4), 457–481. https://doi.org/10.1207/s15427625tcq1504_3.

Sun, H. (2009a). Designing for a dialogic view of interpretation in cross-cultural IT design. In N. Aykin (Ed.), *Internationalization, design and global development* (pp. 108–116). Heidelberg: Springer.

Sun, H. (2009b). Toward a rhetoric of locale: Localizing mobile messaging technology into everyday life. *Journal of Technical Writing and Communication, 39*(3), 245–261. https://doi.org/10.2190/TW.39.3.c.

Sun, H. (2012). *Cross-cultural technology design: Creating culture-sensitive technology for local users*. New York: Oxford University Press.

Taylor, D. (1992). *Global software: Developing applications for the international market*. New York: Springer.

Thayer, A., & Kolko, B. E. (2004). Localization of digital games: The process of blending for the global games market. *Technical Communication, 51*(4), 477–488.

van Reijswoud, V., & de Jager, A. (2011). The role of appropriate ICT in bridging the digital divide. In K. St. Amant & B. A. Olaniran (Eds.), *Globalization and the digital divide*. Amherst: Cambria Press.

Walton, R. (2013a). Civic engagement, information technology, & global contexts. *Connexions: An International Professional Communication Journal, 1*(1), 147–154.

Walton, R. (2013b). How trust and credibility affect technology-based development projects. *Technical Communication Quarterly, 22*(1), 85–102.

Walton, R. (2016). Supporting human dignity and human rights: A call to adopt the first principle of human-centered design. *Journal of Technical Writing and Communication, 46*(4), 402–426. https://doi.org/10.1177/0047281616653496.

Whitney, J. G. (2013). The 2010 citizens clean elections voter education guide: Constructing the "Illegal Immigrant" in the Arizona voter. *Journal of Technical Writing and Communication, 43*(4), 437–455.

CHAPTER 2

Biometric Technology: The Savior of a Risky Electoral System

In the previous chapter, I reviewed the narrow representation of users in localization processes and argued that it is relevant to pay attention to the complex context of use if we are to recover the lost voices of users in localization processes. This means it is important to focus on sociocultural context, space, place, location, and the physical environment, so that technology designed will not be far removed from its context of use. To address the issues I laid out in Chapter 1, we need to develop a deep understanding of our users, their needs, and our complex context of use, Ghana. It must be noted that electoral processes in Ghana are embedded in a complex relationship between politics, economy, history, and local logics.

Once upon a time, Ghana used to be a hub for slave raiders and traders. What used to be the point of no return for Africans who were captured in slave wars and transported to the Caribbean and North America has become an attractive spot for celebrities, presidents, and important dignitaries across the globe. For example, CNN lists Ghana as one of the 19 places to visit in 2019. Ghana came hard at the heels of New Zealand, Egypt, and Japan. In the write-up, CNN Travel introduces the postcolonial country as "West Africa's poster nation for economic success and political stability."[1] More so, in 2013, the *New York Times* listed Accra, the capital of Ghana, as a fourth desirable destination. It is obvious that

---

[1] https://www.cnn.com/travel.

there is a lot to pontificate about this postcolonial country, but in this book I discuss the politics of Ghana and how the introduction of the biometric technology helps to discuss the subjection of elections to digital technology.

Therefore, this chapter provides a historical overview of Ghana with specific attention to the political events that necessitated the adoption of the biometric. For the benefit of my broader audience, I will also provide a brief geographical orientation of Ghana. The use of the biometric technology in Ghana's electoral system indicates how technology adoption and use are embedded in cultural, historical, political, and social relations. It is an indication of how the biometric is situated in the interplay of immediate, community, and sociocultural contexts. It also reveals how technology use is situated in local and global relations, while also speaking to how a local problem is responded to by the use of a global solution: biometric technology.

The political history of Ghana and the brief description of the biometric will help us all to appreciate the global logics which pushed the EC and other stakeholders to adopt this specific technology to solve a local problem. The overview helps me to elucidate the question: How did Ghanaians or the Electoral Commission of Ghana decide to adopt a biometric verification system? We are able to understand how Ghanaians developed a specific need, that is, a need to adopt a biometric verification system to solve specific problems. As a technical rhetor, I say that the overviews provide a rhetorical situation, that is, "the context in which speakers, writers [decision makers, problem solvers] create rhetorical discourse" (Bitzer, 1992, p. 1); the rhetorical situation provides a web of relationships between the various historical, cultural, rhetorical, political, legal, and economic issues that pushed or pulled the country to adopt the biometric technology.

Ghana, like many nations in Africa or many developing democracies, encounter electoral processes that are inhibited by malpractices and various forms of fraud that make it hard for such countries to attain the "genuine" election of representatives as enshrined in the *Universal Declaration of Human Rights*. Discourses about electoral processes in Ghana indicate the extent to which voters have lost confidence in the electoral process and their desire for a more credible system. It is unarguable that the integrity of an electoral process is held high when the public is confident that the processes are free, fair, and incontrovertible. In other words, "the legitimacy of an election and public confidence in the resulting structures of democratic governance largely depend on the actual and perceived integrity of the electoral process. If citizens and candidates believe an election was unfair or poorly-administered, they may not accept the outcome." ("Election Intergrity," 2017), that is, if the electoral system is perceived to be corrupt, or if handlers of elections are perceived to be corrupt or incompetent, the likelihood that results will be accepted in good faith is minimal. Such has been the situation of Ghana. Over the years, political parties and stakeholders have doubted the credibility of electoral results. In most cases, EC officials are perceived to be in bed with or working to favor the government in power. In essence, political parties have been disgruntled about the manner in which EC organizes elections and confidence in the EC has dwindled over the years. As if this is not enough, because these political parties have established media houses, they peddle their perceptions to the public. It is not as though the EC of Ghana has given the voters any course to discredit them. The EC believes that it has done everything possible to win Ghanaian voters over and instill credibility into the electoral process. For instance, it moved from using opaque ballot boxes to transparent ones, and it introduced photo ID cards for all voters around the country to replace the thumbprint ID cards. These efforts, notwithstanding, Ghanaians wanted more to be done.

During my interviews with EC officials, most of them revealed that one of the major issues hurting the integrity of elections in Ghana is the perception that the EC was corrupt. Mr. Gadugah and Mr. Manu, for instance, believe that Ghana's electoral woes can be blamed on the media, political parties, and candidates. More so, Ghana has a high illiterate population, many of whom cannot make informed decisions about who to vote for. Hence, when they hear tales about electoral malpractices and plots to rig, they also spread the false rumors around without pausing to verify.

It may be true that Ghana has a high illiteracy rate or that many people do not have informed knowledge about political activities in the country, but, as Norris (2014) indicates, such individuals can benefit from heuristic cues such as news headlines. Therefore, Norris has argued that it is important to deal with perceptions about election integrity in a country. According to Norris (2014) even if the perception is erroneously conceived, it can still trigger protests and other forms of resistance (p. 92). And indeed, Ghana has seen several of such protests in its electoral history, especially in its fourth republic (1992 to present).

In 1992, for example, the opposition New Patriotic Party (NPP) published the controversial report *The Stolen Verdict* which documented alleged malpractices that inhibited the success of the 1992 elections. To register their displeasure of the electoral process, the party boycotted the 1992 parliament. And in 2008,

> Ghana was at the brink of conflict in the month of December 2008. The NDC complained that it had unveiled a conspiracy between the NPP and the EC to rig the elections. Candidate Prof. Mills warned that a Rwanda-style civil war was imminent. Macho men became part of the process as candidates engaged them. Deaths occurred, several people were maimed. Some Ghanaians looked for their passports and sought solace abroad, as many stored food against the expected "war". Churches and mosques were filled with panic-stricken prayer warriors. As the final results were being awaited, NDC youth armed to the teeth invaded the EC Head Office and burnt vehicle tyres. War was invoked. Rwanda was rehearsed in the city of Accra. In 2012, the panic scenario was repeated. A foreboding doom clouded the hemisphere as once again, the prayer warriors congregated for God's intervention. (Oquaye, 2012, p. ii)

Even Members of Parliament are not hesitant to admit that there are "emerging challenges in the country's electoral system" (*Parliamentary Debates: Official Report [Emergency Meeting]*, 2012, p. 785). A typical example is the statement made by Honorable Mathias Ntow, the NDC member of parliament for Aowin constituency, in parliament during the biometric adoption process that "I am only praying that everything will go on smoothly for this election to be very successful and that at the end of the day, each one of us will be satisfied with the results and accept it in good faith" (p. 795). These short narratives indicate the extent to which stakeholders "question the confidence and competence of the election management body" (*Report of the Commonwealth Observer Group: Ghana Presidential and Parliamentary Elections*, 2013, p. 6) and wished for a system "which will put results beyond dispute" (Oquaye, 2012, p. ii).

Context of Study and the Exigence for Research

Formerly known as the Gold Coast—a name which depicted the country's wealth in gold—Ghana is "a product of a colonial and postcolonial legacy with its historical, socioeconomic, political and cultural problems" (Agboka, 2013, p. 31). Geographically, Ghana lies between latitude 0° N and 23° N. This means that the country lies a little above the equator and within the tropical zone. The implication of this geographical location is that the country is relatively warm or hot all year round. In her book *Culture Smart*, Utley (2009) tells those visiting Ghana for the first time that "Ghana is hot…if you are coming to Ghana, prepare to sweat and take precautions against the sun" (p. 15). Mostly, temperatures range between 21 °C and 40 °C. In addition to the warm conditions, Ghana also experiences a dry weather condition between November and March, a period which is usually referred to as the Harmattan. During this time of the year, dry winds blow from the Sahara Desert across the country. The implication is that the country experiences dusty, humid, and hazy conditions.

The country, by virtue of its geographical location, is bordered by Burkina Faso to the north, Togo to the east, Cote D'Ivoire to the west, and the Gulf of Guinea to the south. Ghana is the only English-speaking country among its neighboring countries which are predominantly French. This close proximity to Francophone neighbors has encouraged some Ghanaians to learn to speak French so they could communicate with neighbors. Also, the educational curriculum of Ghana makes it possible for every Ghanaian to study basic French. For over three decades, the country was divided into 10 regions each with a regional capital: Ashanti (Kumasi), Northern (Tamale), Central (Cape coast), Volta (Ho), Eastern (Koforidua), Brong Ahafo (Sunyani), Upper East (Bolgatanga), Upper West (Wa), Western region (Sekondi-Takoradi), and Greater Accra (Accra). Accra is also the capital of Ghana. However, a referendum on December 27, 2018 saw the creation of six additional new regions. Hence, the country currently has 16 administrative regions, namely Upper West, Upper East, North East, Northern, Savannah, Brong Ahafo, Bono East, Oti, Ahafo, Ashanti, Eastern, Western North, Western, Central, Greater Accra, and Volta. These regions are inhabited by people with varied migration stories and culture. For instance, the Ashanti region is mostly populated by a variant of Akan ethnic group known as *Asantes*, the Volta region by *Ewes*, the Central region is inhabited by the *Fantes*, another variant of Akan. In spite of these unique identities, these regions are not inhabited by individual ethnic groups. This is to say that Ashanti region is not inhabited by Asantes only but amalgamation of people from other regions in Ghana. People have intermarried as well, so it is a cliché or ethnocentric to link one cultural group to a specific region in Ghana.

The country's strategic geographical location makes it an attractive hub for commerce. With agriculture as the major source of employment for most of the citizenry (Smith, 2012, p. 12), the country's economy is and has always relied on the exportation of cocoa, gold, timber, diamond, manganese and, more recently, crude oil. The country's educational system is, as Smith (2012) identifies, "a remnant of colonial domination" with emphasis placed on "alphabetic literacies—reading, writing and computing" (p. 19).

Ghana, a former colony of Britain, became the first country in sub-Saharan African to gain independence on March 6, 1957 and a republic on the of July 1, 1960. In its 59 years after independence, Ghana has witnessed four major military upheavals in 1966, 1972, 1979, and 1981. In 1992, the country began another road to democracy by adopting a new constitution that established the fourth republic in 1993. Ever since the fourth republic was established, Ghana has organized six successful elections in 1996, 2000, 2004, 2008, 2012, and 2016, respectively. In 2000, Ghana voted into power the opposition leader, John Agyekum Kufuor and the NPP. This was a monumental occurrence because it was the first time a ruling party was handing over power to an opposition leader in the fourth republic, but this has in a way become the norm since in 2008, the NPP lost power and handed over to the NDC led by Professor John Evans Atta Mills; in 2016, the NDC lost the general elections and subsequently handed over power to the NPP. I am, however, careful with the use of "successful" elections because the history of elections in Ghana has been fraught with such electoral malpractices as over voting, bloated voters' register, and vote rigging.

In spite of these incongruities, Ghana has been hailed as the beacon of democracy in Africa. For instance, when Barack Obama made his first historic visit to Ghana as the president of the US, he described Ghana as a "model democracy," and on his visit to the country in 1998, president Bill Clinton referred to Ghana as "the gateway to Africa." Apart from these references, Ghana is also noted for its political stability. At least the country has to an extent managed to avoid post-election violence—an achievement which sets it apart from nations such as Cote D'Ivoire, Liberia, Zimbabwe, Kenya, Togo, and Benin which have seen post-election conflicts. It is also not surprising that the national football team of Ghana bears the sobriquet 'the Black Star.' It speaks of the role of Ghana in the affairs of the African Continent—As the first country in Sub-Sahara Africa to gain independence, Ghana has been at the forefront of conversations about the integration of the African continent. On the night of independence, Dr. Kwame Nkrumah, the leader of the Conventions People Party (CPP), the party which gained independence for the country, and the first president of the country intimated that the independence of Ghana was meaningless unless it was linked to the total liberation of Africa. The zeal to move toward decolonialism led Kwame Nkrumah to coin the word "neocolonialism," to resist all forms of exploitation or further control of Africans by colonial masters through indirect means. Ghana was instrumental in the formation of the Organization of African Unity (now African Union) in 1963; A Ghanaian judge, Akua Kuenyehia, serves as a judge at the International Criminal Court; and more recently, Ghana produced the seventh Secretary-General of the United Nations. I am speaking specifically of Kofi Annan. Ghanaian soldiers have served on UN Peace Missions in Lebanon and DR Congo; and the Electoral Commission of Ghana has mentored Electoral Commissioners from other parts of the African continent.

Ghana, thus, is fighting a myriad of battles: First, it needs to make sure its electoral system does not degenerate into any major civil war. In this regard, it has to make sure it sustains confidence in the electoral process; and second, its well-acclaimed title "star of Africa" must be guarded and communicated to the external world by every means possible. Citizens cannot be let down; neither should external institutions which have placed their hopes in the country be let down.

Vote Rigging, Machoism, Ballot Snatching, and the Fear Factor

Several scholars have documented electoral malpractices that have characterized the country's political terrain. For instance, Heinz Jockers, Dirk Kohnert, and Paul Nugent identified in "The Successful Elections of 2008: A Convenient Myth?" that block voting in the Ashanti and Volta Regions in Ghana influenced "the level of electoral malpractice designed to stack up additional votes in areas of strength" (Jockers, Kohnert, & Nugent, 2010, p. 97). In their article, they conjectured that "vote rigging…could acquire a decisive importance, possibly representing an even more important 'third force'" (Jockers et al., 2010) than smaller parties in Ghana. They also identified issues of inflated voters' registration which consequently resulted in high voter turnout, especially in the Ashanti and the Volta regions—the strongholds of the NPP and the National Democratic Congress (NDC), respectively. These and many other issues make the authors conclude that the electoral success of Ghana is a convenient myth.

In another study, Smith (2012) reported that prior to the 2000 elections, the electoral commissioner, Dr. Afari-Gyan "stated publicly that there was a substantial 'bloating' of the register..." (p. 624). Interestingly, though the provisional national census figures recognized 18.4 million Ghanaians of which half was of voting age, the Electoral Commission had recorded 10.7 million names on the voters register "roughly 58% of the country's total population" (ibid., p. 624). Per the census data, the author identified that the "EC's register was bloated with more than 1.5 million ghost voters" (ibid., p. 624). Smith and Jockers et al. are not the only scholars who have identified such malpractices as over voting and bloated register: Oquaye (1995), Saaka (1997), Salia (2000), and Gyimah-Boadi (2004) have also uncovered these in their scholarship. Added to these malpractices is the feeling of fear and insecurity that pervades the atmosphere during electoral seasons. This feeling of insecurity of the polls reached its magnitude when in 2008 Ghana came close to experiencing post-election conflict.

The malpractices enumerated in the preceding paragraph; vote rigging, over voting, bloated register, and the use of thugs to harass voters are not peculiar to Ghana. Cheeseman (2015), for instance, indicates with several examples that democracy or elections in Africa has been characterized by vote rigging, intimidation of opposition leaders, intimidation of voters, electoral violence, disappearance, and accusations of electoral manipulation. He identifies that these "democratic breakdowns...continue to be a prominent feature of multiparty politics up to the present day" (p. 1). But he is not quick to give up on the future of democracy in Africa. He argues that the sources of the electoral malpractices are multivariate and multilayered, and they have traditional and historical antecedents. He identifies, and I agree to an extent, that the causes can be linked to: the rapid nature with which countries are introduced to multiparty elections, coup d'états, colonial rule, and the attainment of independence. There are also such underlying issues as neo-patrimonial rule, gate-keeping, and mixed feelings about nationalist struggles.

As an example, I state how the first two instances have led to electoral violence or vote rigging. Nic Cheeseman indicates that the idea of elections or multiparty democracy is introduced or reintroduced after a country has witnessed a form of violence or after a country has experienced one-party rule for a number of years. In these instances, warring factions find it difficult to trust and commit to the idea of elections. In such countries, elections only exacerbated crisis because it "increased the level of political uncertainty in fragile and divided societies" (p. 144). He cites cases in Angola (civil war in 1993), Rwanda (1994 genocide), and Cote D'Ivoire (civil war in 2000) as instances where violence could be linked to the introduction of party politics. In most cases, calls for the reintroduction of elections or multiparty politics are not made by locals but instead are orchestrated by such foreign donors as USAID, IMF, European and North American governments, and even the World Bank. In other cases, democracy was necessitated by both domestic and international stakeholders. In cases where these donors pushed governments to reintroduce elections, they could not adequately shape processes that could reconstruct nations and create a more democratic system (p. 115); hence, such countries enter the abyss of perpetual "infrastructural weaknesses within the government and individual political parties…" (p. 35).

Another root cause of electoral malpractice on the African continent, Cheeseman reveals, can be linked to the numerous coup d'états that have plagued the continent. It is indisputable to state that more than half of the countries in Africa have seen instances of military rule. As I noted above, Ghana has witnessed four military rules; in 1963, Benin, Congo-Brazzaville, and Togo experienced military junta; in 1966, Central African Republic witnessed a coup d'état; Guinea Bissau and Nigeria are among the several countries that have witnessed military take-overs. The logic of military intervention is summed up by Cheeseman (2015) as (1) the rhetorical claim made by military personnel that they want to improve on living standards; and (2) military intervention as the solution to a gap or chasm created by political instability in a country (p. 44). Military interventions in Ghana, to an extent, exhibit the trait of the first logic identified by Nick Cheeseman. For instance, when Jerry Rawlings staged a coup for the second time in 1981, he stated: "Fellow citizens of Ghana,…I ask for nothing less than a Revolution, something that would transform the social and economic order of this country…." He went on to propose an approach to governance which is "designed to make the people take their destiny into their own hands and feel they are part of the government and not it being something above them" (quoted in Oquaye, 2004, p. 97) and in 1972 when Ignatius Kutu Acheampong ousted Abrefa Busia, he hinted he was going to provide "food on the table" (ibid., p. 58).

The interesting thing about these coup leaders is that most of them ended up forming political parties so they could consolidate their position or put a democratic face to their autocratic regimes. Jerry Rawlings who entered the political scene of Ghana in 1979 and 1981 later formed the NDC for the 1992 general election. In order to maintain power, most of these military-leaders-turned-democrats orchestrated several dubious plans to win elections. To this end, those who opposed military leaders such as Rawlings and his PNDC cohorts were "perpetually harassed by the security agencies" (Oquaye, 2004, p. 488). The spirit to resist military authoritarianism by pro-democrats, on the one hand, and the desire of authoritarian, military rulers to hold on to power, on the other hand, introduced violence, electoral rigging, machoism witchcraft, and other negative malpractices into the political sphere of many countries in Africa (Cheeseman, 2015; Oquaye, 2004). These practices, I believe, are adopted so more votes could be gathered or reduce the votes of opposing parties: It is a strategy to gain power and control of the resources because whoever wins power gets to execute influence over society, protect the interests of their close-knit networks, and "establish and sustain political hegemony" (Cheeseman, 2015, p. 26).

Biometric to the Rescue

Amidst the quagmire and rumors of election rigging, Ghana, spearheaded by the Electoral Commission, adopted the biometric verification voting system for the 2012 presidential and parliamentary elections. The Electoral Commission, the constitutionally mandated institution charged, among other things, to conduct and supervise all public elections and referenda in Ghana (*1992 Constitution of the Republic of Ghana*, 1992, p. 39), asserted that the device will:

- assist in detecting and preventing practices of impersonation and multiple voting
- "expose" electoral offenders
- "provide transparency" in results

And in the Parliament of Ghana, the lawmakers also made a case that the biometric will "make it extremely difficult for a person to use the name and particulars of another voter." Therefore, the most popular phrase in

Ghana before, during and after the 2012 election was "No Verification No Vote," a phrase which indicates the important role the biometric technology will play in the electoral process. Hence, by adopting the biometric device, the country's electoral system was subjected and "articulated to the logic of [a] digital technology" (Hristova, 2014, p. 516). On a surface level, I argue that these claims are made to establish the ethos and the neutrality of the biometric technology. To an extent, through these narratives we encounter "how to behave and what to think, feel, fear, and desire—and what not to" (Kitalong, 2000, p. 291) about the biometric technology.

Frankly speaking, most African countries like Ghana are working tirelessly to make electoral processes "clean" and violence-free and instill confidence in the electoral system. In Africa alone, countries such as Ghana, Kenya, Nigeria, and Cameroon have initiated such moves (Makulilo, 2017, p. 202) because of the level of trust surrounding the technology. For these countries, biometric is used to prevent two major issues that bedevil free and fair elections: double registration and multiple voting. It is the earnest desire to see success in the electoral system that the Electoral Commission of Ghana and stakeholders decided to introduce biometric technology to curb electoral malpractices. So, when the conversation on adopting a biometric system began in 2010, I became very interested and wondered how clean the electoral process was going to be when the country finally adopted it. At that point, I shared the assumption of many others that technologies were the panacea to social woes. I became more motivated to embark on this study when after the 2012 elections, the opposition NPP filed a petition to the Supreme Court of Ghana contesting the validity of the election of John Dramani Mahama as the president of the Republic of Ghana. According to the NPP, there were the recurrence of such issues as over voting, impersonation, vote rigging, and many other malpractices that characterized the 2012 elections (Gyimah-Boadi, 2013; Kelly & Bening, 2013; MyJoyOnline, 2013; Oquaye, 2012; *Supreme Court Judgement on the Presidential Election Petition, Akufo-Addo V. Mahama and Others*, A.D 2013). The then Chairman of the Electoral Commission, Dr. Afari-Gyan, admitted at the Supreme Court during the election petition that under the condition the commission operated, "...administrative and clerical errors are unavoidable" (Baneseh, 2015, p. 25).

Based on the challenges enumerated above, it is without a doubt that the electoral process of Ghana is at risk. Defining the electoral process as at risk or the perception that elections in Ghana had experienced crisis moments provided a *kairotic* moment for the introduction of the biometric; hence, the biometric provided a kairotic response to our risky electoral process. The biometric, it was hoped, would provide a moment of change in the electoral process of Ghana. The biometric will provide intervention in the sense that it will make it "extremely difficult for a person to use the name and other particulars of another person to vote" (p. 788). Defining the moment, in this regard, also helps to identify the cause of the electoral woes of the country: people vote more than once. Thus, with the biometric, the country will, *ipso facto*, seize the moment, it will go through a credible and reliable electoral process that will yield accurate results. In this regard, the biometric becomes the "saving grace" of our electoral process. Why did the EC and stakeholders choose the biometric and not a different technology?

Biometric Technology: A Global Response to a Local Problem?

Biometric technologies involve

> The collection with a sensoring device of digital representations of physiological features unique to an individual, like a fingerprint…This digital representation of biometric data is then usually transformed via some algorithm to produce a so called [sic] 'template…These templates are stored in a centralized database that is accessed when on following occasions the finger…is presented to the system. After a similar algorithm transformation of this second image, a comparison can be executed. If a matching template is found, the person presenting themselves is 'recognised' and counts as 'known' to the system. (cited in Hobbis & Hobbis, 2017, p. 117)

Because the device captures the biological and physical features of individuals through the use of computers, scanners, and webcams, it is believed that the biometric collects accurate information about individual so that "when we say it is you, we know it is you" (stlghana, 2012). As a result of this perceived trust that biometric captures accurately, the technology has become an integral part of our globalized, interconnected world; it is "used by government agencies and private industry to verify a person's identity, secure the nation's borders,...to restrict access to secure sites including buildings and computer networks," (Vacca, 2007, p. 3) and it has become nearly impossible to travel from one country to another without a biometric passport. The primary function of a biometric machine is to authenticate and verify individuals. This function is explicitly stated in the foreword to the user manual: The device is constantly referred to as a "verification device." Biometric verifies by scanning "a subject's physiological, chemical or behavioral characteristics in order to reify or authenticate their identity" (Pugliese, 2010, p. 2). These technologies make use of devices such as scanners and cameras to "capture images, recordings, or measurements of an individual's characteristics and use computer hardware and software to extract, encode, store and compare these characteristics" (Vacca, 2007, p. 23). Biometric technology is used to "verify a person's identity, secure the nation's borders, and to restrict access to secure sites..." (Vacca, 2007, p. 3). Mostly, biometric is used to identify "you as you" (Pugliese, 2010) and also "holds the promise of fast, easy-to-use, accurate, reliable, and less expensive authentication for a variety" (Vacca, 2007, p. 15). The biometric system, per its design, answers two main questions that relate to identification and verification:

- Verification: It is used "to verify a person's identity". This is done in order "to authenticate that individuals are who they say they are" (Vacca, 2007, p. 23). Am I who I say I am? (Magnet, 2011)
- Identification: It is used to establish a person's identity—here, the system determines who a person really is. It responds to the question: Who am I? (Magnet, 2011; Pugliese, 2010).

Biometric scholars indicate that the technology functions on three levels or stages. First is the enrollment stage. The subject presents himself/herself in order to enroll in the system. This process of presenting "allows biometric technology to capture an image of the particular imprint of the subject" (Pugliese, 2010, p. 2). One must provide a form of identification in the form of an ID. Then, you present the biometric finger. When the finger is presented, "distinctive features located on one or more samples are extracted, encoded and stored as a reference template" (Vacca, 2007, p. 22). Thus, the first stage acquires information about your physical behavioral traits (Magnet, 2011, p. 21).

The second stage is the verification stage. The captured image is digitally converted through the use of particular algorithms into a template (Magnet, 2011; Pugliese, 2010; Vacca, 2007). Your enrolled body must be verified by the system. This way, the "system compares the trial biometric template with this person's reference template which is stored in the system" (Vacca, 2007, p. 23). This is referred to as 1:1 matching, meaning: there can be only you and no one.

The final stage sees to it that "the trail template is compared against the stored reference templates of all individuals enrolled in the system" (Vacca, 2007, p. 25). A person's physiological and behavioral characteristics are turned into a "digital representation in the form of a template" (Magnet, 2011, p. 21). Thus by adopting the biometric, the nation declared the elections a security concern which needed to be addressed. Voters became bodies that must be controlled and disciplined. These processes ensure that the biometric technology will capture accurate information and release accurate results.

The telos of the biometric, therefore, is to ensure transparency, effectiveness, and accuracy in the electioneering process. This idea that the biometric will ensure or restore credibility to the electoral process which witnesses irregularities follow a specific logic: Biometric technologies are neutral, efficient, accurate, honest, and error free. I refer to this logic as biometric rhetoric. Hobbis and Hobbis (2017) indicate that the air of "certainty" surrounding biometric technologies has propelled "the rapid, global expansion of the biometric industry" (p. 117). Considering the accelerated nature of its adoption and use, market analyst have predicted an expansion of the technologies market size from US$ 9.58 billion in 2015 to US$ 31 billion in 2023 (cited in Hobbis & Hobbis, p. 117). As the scholars cited here indicate, 34 low-to-middle-income countries have opted to utilize the biometric to enhance their electoral system. Biometric technology, thus, provides hope for countries that seek to clean their electoral system because of the belief that it is transparent and authentic and captures subjects accurately.

I will reveal that noble intentions to adopt a technology to resolve social issues can turn sour if stakeholders and designers do not work to properly localize the technology.

REFERENCES

1992 Constitution of the Republic of Ghana. (1992).
Agboka, G. Y. (2013). Participatory localization: A social justice approach to navigating unenfranchised/disenfranchised cultural sites. *Technical Communication Quarterly, 22*(1), 28–49. https://doi.org/10.1080/10572252.2013.730966.
Baneseh, M. A. (2015). *Pink sheet: The story of Ghana's presidential election as told in the Daily Graphic.* Accra: G-Pak Limited.
Bitzer, L. F. (1992). The rhetorical situation. *Philosophy & Rhetoric, 25,* 1–14.
Cheeseman, N. (2015). *Democracy in Africa: Successes, failures, and the struggle for political reform* (Vol. 9). Cambridge: Cambridge University Press.
Election Integrity. (2017). *International foundation for electoral systems.* Retrieved August 26, 2017, from http://www.ifes.org/issues/electoral-integrity.
Gyimah-Boadi, E. (2004). *Ensuring violence-free December 2004 elections in Ghana: Early warning facilities.* Paper presented at the Briefing Paper, Ghana Center for Democratic Development.
Gyimah-Boadi, E. (2013). *Strengthening democratic governance in Ghana: Proposals for intervention and reform.* Accra: A Publication of Star Ghana.
Hobbis, S. K., & Hobbis, G. (2017). *Voter integrity, trust and the promise of digital technologies: Biometric voter registration in Solomon Islands.* Paper presented at the Anthropological Forum.
Hristova, S. (2014). Recognizing friend and foe: Biometrics, veridiction, and the Iraq War. *Surveillance & Society, 12*(4), 516.
Jockers, H., Kohnert, D., & Nugent, P. (2010). The successful ghana election of 2008: A convenient myth? *The Journal of Modern African Studies, 48*(01), 95–115.
Kelly, B., & Bening, R. (2013). The Ghanaian elections of 2012. *Review of African Political Economy, 40*(137), 475–484.
Kitalong, K. S. (2000). "You will": Technology, magic, and the cultural contexts of technical communication. *Journal of Business and Technical Communication, 14*(3), 289–314.
Magnet, S. (2011). *When biometrics fail: Gender, race, and the technology of identity.* Durham, NC: Duke University Press.
Makulilo, A. (2017). Rebooting democracy? Political data mining and biometric voter registration in Africa. *Information & Communications Technology Law, 26*(2), 198–212.
MyJoyOnline. (2013). Election 2012 petition hearing-day 2. *Election 2012 petition hearing.*
Norris, P. (2014). *Why electoral integrity matters.* New York: Cambridge University Press.
Oquaye, M. (1995). The ghanaian elections of 1992—A dissenting view. *African Affairs, 94*(375), 259–275.

Oquaye, M. (2004). *Politics in Ghana, 1982–1992: Rawlings, revolution, and populist democracy*. Accra and New Delhi: Tornado Publications.

Oquaye, M. (2012). *Strengthening Ghana's electoral system: A precondition for stability and development*. IEA Monograph (No. 38). Ghana.

Parliamentary Debates: Official Report (Emergency Meeting). (2012). Retrieved from Parliament House, Accra.

Pugliese, J. (2010). *Biometrics: Bodies, technologies, biopolitics* (Vol. 12). New York: Routledge.

Report of the Commonwealth Observer Group: Ghana Presidential and Parliamentary Elections. (2013). Retrieved from London. http://thecommonwealth.org/sites/default/files/inline/GhanaElections-FinalReport2012.pdf.

Saaka, Y. (1997). Legitimizing the illegitimate: The 1992 presidential elections as a prelude to Ghana's fourth republic. *Issues and trends in contemporary African politics* (pp. 143–172). New York: Peter Lang.

Salia, A. (2000). Voters Register in excess of 1.5 m people—Afare-Gyan. *Daily Graphic, 21*, 2000.

Smith, B. (2012). *Reading and writing in the global workplace: Gender, literacy, and outsourcing in Ghana*. Lanham: Lexington Books.

stlghana. (2012). *STL Ghana—Biometric system voter registration process*.

Supreme Court Judgement on the Presidential Election Petition, Akufo-Addo V. Mahama and Others (A.D 2013). Retrieved from http://www.myjoyonline.com/docs/full-sc-judgement.pdf.

Utley, I. (2009). *The essential guide to customs and culture: Ghana*. London: Kuperard.

Vacca, J. R. (2007). *Biometric technologies and verification systems*. Burlington, MA: Butterworth-Heinemann.

CHAPTER 3

Decolonial Methodology as a Framework for Localization and Social Justice Study in Resource-Mismanaged Context

I have indicated from the previous chapter the complex nature of Ghana's political terrain and the multifaceted issues that encumber successful elections in the country. To this end, a fruitful study of why the technology was adopted and the role technical documents played in the adoption process requires a complex methodology—a methodology that is sensitive to the political history of Ghana—history of electoral violence, efforts made to prevent the malpractices that have bedeviled successful elections; a methodology which can capture local knowledge constructed about the biometric technology and the various attempts made to localize the biometric to fit the complex context of use. I believe a decolonial framework is an appropriate methodology to study the situations I have outlined. In this chapter, I will demonstrate the relevance of decolonial methodology to localization and social justice projects in resource-mismanaged or intercultural/international contexts.

I see research methodology as a critical and flexible heuristic which is rhetorically constructed by researchers to meet the object of study. My position is that "all methodology is rhetorical, an explicit or implicit theory of human relations which guides the operation of methods" (Sullivan & Porter, 1997, p. 11). This position admits that a research methodology should be critical, reflective, situated, and rhetorical; it should acknowledge the dynamic relationship between rhetoric and culture, provide avenues for intervention, and recognize research as ethical and political. These principles become more relevant when one is researching international contexts which have complex rhetorical cultural situations.

International technical communication scholars have identified the complexity in studying international, intercultural, or cross-cultural context and have advocated the use of methodologies that can capture the complex nature of the context under study. Yet, scholars are still grappling to develop a systematic methodology to guide research conducted in this area of scholarship. Ding (2014) aptly states that systematic approaches to the study of professional communication in international, cross-cultural contexts does not exist, and Thatcher (2010) also observes that even though globalization has influenced the work of professional communicators, scholars have still not developed any viable approach to the study of global rhetoric in professional contexts. Nonetheless, the varied approaches to international/intercultural technical communication research are certain of one thing: Culture is the most important factor to consider (Ding & Savage, 2013; Lovitt & Goswami, 1999; Sun, 2004, 2006, 2012; Thrush, 1993; Walton, 2014), but it has become nearly impossible to define culture or capture the complexity of the term. Attempts to do so have led to either a narrow discussion of culture or over-generalization, oversimplification or stereotypical and static representation (Agboka, 2014; Ding, 2014; Sun, 2012). For example, Sun (2012) identifies that approaches employed to study culture capture local cultures in terms of do's and don'ts, anecdotes and business cases which only report the dominant culture in a country; Agboka (2012) states that intercultural approaches have favored "'large culture' ideologies" (p. 159); and Ding (2014) recounts how the focus on nation-states as the unit of analysis limits the analytical tools used to capture rhetorical and communicative practices. She also recounts how approaches rely on oversimplified notions of cultural variables and cultural dimensions (p. 14) or that they are preoccupied with satisfying the needs of industries (p. 15). Barry Thatcher identifies that the field still, unabashedly, relies on critical and cultural approaches that favor local over global. This overreliance on antiquated approaches presents such problems as "ethnocentrism, methodological aporia, poorly theorized global-local relations, ignoring large scale variables… and unworkable ethics" (p. 2). In essence, scholars have constructed a hegemonic notion of culture to create a social control among cultural elements. And this means that we get to understand culture from a privileged standpoint to the neglect of other ways of seeing culture.

In order to respond to the complex nature of intercultural contexts and also to fruitfully conduct their research, scholars have developed different heuristic approaches. For example, Thatcher (2011) develops an intercultural rhetorical model which emphasizes the need for research in international contexts to capture and connect how cultures conceive of or define the self, their epistemologies, ideologies, or and how these three categories link up with rhetorical patterns (pp. 3–5). His approach stresses the need to understand how rhetoric and culture interact with or "act upon" each other in intercultural contexts. Sun (2012) develops *Culturally Localized User Experience* (CLUE) methodology to study how users localized and integrated mobile technology into their daily lives; Ding develops *Critical contextualized Methodology* as an analytic tool to study the various ways a global epidemic such as the Severe Acute Respiratory Syndrome (SARS) was constructed both rhetorically and culturally by China, the US, Canada, the World Health Organization, and other major stakeholders. This study enables her to advance a rhetoric of global epidemics which focus on the communication of global epidemics and the ways that political, ideological, and institutional structures shape the framing of global epidemics.

Decolonial Methodology Defined

It is important to acknowledge that these numerous approaches are developed by scholars who live, work, attend schools, and perform research in Western contexts, and as a result of their relationship to the West, the tendency that their approaches will follow the Western scientific approach to conducting research is high. In essence, these approaches may carry the stigmata of colonialism which will not be adequate to study post-colonial contexts. For instance, Linda Tuhiwai Smith (2012), a major proponent of decolonial methodologies, identifies that colonized communities are more skeptical of Western researchers and their approaches because to the colonized, the word "'research' is inextricably linked to European imperialism and colonialism" and when this word is mentioned in colonized contexts, "it stirs up silence, it conjures up bad memories, it raises a smile that is knowing and distrustful" (p. 1). The colonized have reservations about research because it is the tool that drives imperialist motives and "research" has been used in the time past as a tool to "extract and claim ownership of" indigenous knowledge, imagery, their modes of creation, production and "simultaneously reject the people who created and developed ideas" (p. 1). So have Western theories been used to oppress indigenous people

(p. 31). More so, "the collective memory of imperialism has been perpetuated through the ways in which knowledge about indigenous peoples was collected, classified and then represented in various ways back to the West, and then, through the eyes of the West, back to those who have been colonized" (p. 1).

The implication is that when researchers from the West use Western theories and approaches to study colonized contexts, such researchers recast colonialism and imperialism, and in most cases, local ways of knowing are subjugated, silenced, or not acknowledged. It is a recipe for domination. When Western concepts meet post-colonial contexts, what we see is an ideological tug of war or a contact zone which inadvertently favors Western ideas over non-Western local ways of encountering the world. For instance, Sun (2012) profoundly identifies that even though the term "culture" is very relevant to conversations about technology design in cross-cultural, international contexts, it has been narrowly conceptualized. She also criticizes the approaches used to collect data on culture of championing monolithic values. She identifies that too often, approaches employed to study culture capture local cultures in terms of (1) do's and don'ts, anecdotes, and business cases. This focus, she identifies, only reports the dominant culture in a country; (2) value-neutral cultural dimensions. This approach also promotes the spirit of positivism because culture from this perspective places concrete cultural realities into static dimensions; and (3) structured fieldwork methods which focus on richness and texture of everyday life and is concerned with "the production of and exchange meanings between the members of a society, or a group in an ethnomethodological sense" (p. 14).

Also, Agboka (2014) indicates that the pluralistic approaches advanced by scholars in our field to study international contexts are positive signs that speak to the fields ability to adapt methodologies to meet local needs, yet these approaches are inefficient in capturing complex situations, such as local logics, histories, ideologies, politics, economy, and legal matters in post-colonial contexts (p. 298). He therefore proposes a decolonial methodology which pays attention to "local logics, rhetorics, histories, philosophies and politics" (p. 298) and calls on other researchers to implement decolonial approaches in studying complex international, intercultural, and cultural contexts. According to Agboka (2014), decolonial methodologies enable researchers to be flexible, reflexive, humble, and respectful when approaching intercultural contexts (p. 316); it helps to reveal "the ways that colonialism continues to operate and to affect lives in new and innovative ways as well as to show the unmitigated damage inflicted by past colonial practices" (p. 298), and it offers us a humanistic heuristic for inserting ourselves into research contexts, and then working on research projects that benefit both researcher and participant (p. 304). Haas also indicates that decolonial approaches help to (1) address how colonialism shapes the way we perceive "people, literacy, culture, and community and the relationship therein," and (2) "support the coexistence of cultures, languages, literacies, memories, histories, places, and spaces—and encourage respectful and reciprocal dialogue between and across them" (p. 297). Because decolonial methodologies provide a good alternative to study colonized contexts, I adopt this methodology, and in this chapter I will discuss how decolonial methodology provides a good tool to recovering the lost voices of users in localization and design processes.

At the basic level, decolonial methodology is a recovery process. This methodology seeks to recover the lost identities of colonized people by championing self-determination, empowerment, decolonization, and social justice. It is relevant for a decolonialist to realize that "indigenous ways of knowing were excluded and marginalized" (Smith, 2012, p. 71) and then initiate a move to reclaim, reconnect, and reorder "those ways of knowing which were submerged, hidden or driven underground" (p. 72). A decolonial project must acknowledge that writing and storytelling have been used as tools of subjugation, and so it must aim to "write back, talk back" (Smith, 2012, p. 8); "re-write and re-right their position" (ibid., p. 29) and "talk back to or talk up to power" (ibid., p. 226). Decolonization helps to tell alternative stories from the perspective of the colonized. Decolonial approaches maintain that "indigenous populations share experiences as peoples who have been subjected to the colonization of their lands and cultures, and the denial of their sovereignty, by a colonizing society that has come to dominate and determine the shape and quality of their lives even after it has formally pulled out" (Smith, 2012, p. 7). According to Linda Tuhiwai Smith, decolonization "is a process which engages with imperialism and colonization at multiple levels" (p. 21). To this scholar, a decolonialist does not take such words as "imperialism, history, writing and theory" lightly as these words draw attention to the various ways in "which indigenous languages, knowledges and cultures have been silenced or misrepresented, ridiculed or condemned" (p. 21). As a result of these forms of injustice, subjugation, and domination perpetrated by Western culture, individuals in colonized contexts "perceive a need to 'decolonize our minds' to recover ourselves, to claim a space in which to develop a sense of authentic humanity" (p. 24). Linda Tuhiwai Smith identifies that to decolonize two things need to be entailed: (1) thinking about the notion of authenticity, that is, precolonial moments when colonized communities "were intact as indigenous peoples," the time when natives had absolute control of their lives, and (2) "an analysis of how we were colonized, of what that has meant in terms of our immediate past and what it means for our present" (p. 25).

In sum, decolonial approaches recognize:

1. The ever-increasing commitment to the recognition and realization of social justice
2. Equity and equality for all peoples, underpinned by social models of difference
3. Enhanced sensitivity to the role of discourse in constructing and framing identities and relationships
4. Various consequences of globalization and improved communications and technologies which have had the effect of shrinking the world and bringing people from far-flung places into closer contact with each other (cited in Agboka, 2014, p. 303).

What Does This All Mean?

Before I articulate what all these discussions of decolonial approaches mean for localization and social justice project in international technical communication research contexts, let me emphasize that localization objectives and social justice are interconnected issues, because localization and social justice both thrive to empower the marginalized, silenced, subjugated, and the disenfranchised in the society and in technology design processes. Although it is not explicitly stated, localization experts in technical communication share the same concerns as social justice researchers. They both strive to empower the marginalized and the disenfranchised in the society. To be specific, localization scholars in TPC argue against a top-down approach to technology design. They contend that technologies break down because developers do not take into consideration the culture and nature of use in user contexts. Consequently, designs do not reflect the ideology of users, but instead it only makes developer ideologies visible. To this end, localization experts have sought for ways to help designers to understand that users also have forms of knowledge which must not be marginalized in design processes (Acharya, 2017; Gonzales, 2018; Gonzales & Zantjer, 2015; Sun, 2006, 2009, 2012). In the same way, social justice seeks to empower the marginalized or disenfranchised individuals in our societies or it thrives on equity and meaningful access. Young (1990), as an example, advises that social justice advocates should work to identify and interrogate domination and oppression (p. 16) in our various institutions. She argues that oppression can come in varied forms: exploitation, marginalization, powerlessness, cultural imperialism, and violence (pp. 48–62).

One would see why decolonial methodology is relevant to this project. First, the research context, Ghana, is a decolonial context which has suffered from the shackles of colonialism. As I indicated in Chapter 2, Ghana is still struggling to shake off and properly manage some of the colonial institutions, such as democracy and education. More so, Ghana is still struggling to assert control over some of its indigenous institutions such as culture, local ways of managing affairs, and the economy. When Ghana lost its sovereignty to colonial powers, it lost its local institutions such as the chieftaincy institutions and local forms of socializing the young. Local music has lost its proper place in the Ghanaian society and the extended family institution which establishes Ghana as a Collectivist society is gradually losing its grip. The pride Ghanaians have in saying "(y3w)," meaning "we have" is giving into individualist statements like "(mew)" meaning "I have" and a Ghanaian who seeks to think Ghanaian is labeled "colo" implying that the individual is old fashioned or uncivilized. Moreover, many Ghanaians would love to shop for things made outside Ghana than items locally manufactured. To the average Ghanaian, things made in Ghana are inferior to foreign goods. Many Ghanaians will also celebrate Western events like "the royal wedding" than celebrate a traditional ruler getting married; "Black Friday" than traditional festivals; watch English Premier League (EPL), Spanish La Liga, or German Bundesliga than watch the local football (soccer) league because they find these foreign leagues more attractive. Most worryingly, even though the official currency of the country Ghana Cedi, most Ghanaians prefer to quote items/prices in US dollars than in Ghana Cedi—a phenomenon some economists have labeled the "dollarization of the cedi." These examples indicate how Ghanaian ideologies have been subsumed to foreign ways of thinking and doing things, while indigenous intellectual systems and local ways of doing things are denigrated. The Ghanaian, from decolonial perspective, has lost his/her authentic nature, and so it is necessary to work to decolonize the minds of people.

Second and more important, decolonial helps to acknowledge that users of technology have been marginalized in a lot of ways and that we need to recover their voices. I indicated earlier that decolonial methodologies seek to recover, so do localization and social justice theories. Like the colonized societies, Johnson (1998) describes users knowledge of practice as "a colonized knowledge ruled by the technology and the 'Experts' who have developed the technologies" (p. 5). Here, it is realized that colonialism is not only the physical annexation of a colony or nation, but it also emphasizes the absence of representation. Meaning that if one's voice is not acknowledged in a technology he/she will use, that person's voice has been colonized by designers. Colonialism, I would say, can be physical, psychological, and cultural. As Comaroff and Comaroff (1991) argued, "The essence of colonisation inheres less to political overrule than in seizing and transforming others by the very act of conceptualizing, inscribing, and interacting with them in terms not of their own choosing ..." (p. 15). Hence, although users have not been physically brutalized or physically relocated to new-found worlds, as in the case of colonialism and slavery, they have been silenced and their ways of knowing have been subjugated by expert or colonial knowledge; designers have colonized their voices and identity, neglected and negated their language and culture of use. Therefore, it is necessary to reveal the various ways that users have been delegitimized and their local ways of knowing unappreciated.

I could hear a designer say they do study and collect viable information about users through rigorous scientific research. That is a worthy venture, but, as Smith (2012) tells us, scientific research is an imperialist tool as it is used to extract and perpetuate colonizing agenda. Because designers collect data about users, they may think that they know all about the users, their culture, and ways of knowing, but actually in fact they do not, because the tools they use to collect data are limited and ethnocentric, and as Sun (2012) indicates, users know their contexts more than designers. The tools and mode of data designers used to collect information about users need to be "decolonized," that is, the questions asked about users and how they write about/report on users must be "considered carefully and critically before being applied" (Smith, 2012, p. 41). To this end, decolonial methodology is good for localization as the goal of localization is to help decolonize the myth that users are passive users of technology and present a more positive image of the user as knowledgeable, creative, and an active person who initiates strategies to save technologies.

It is important to mention that although decolonial methodology proposes a decolonization of Western research methods, it does not mean Western ways of constructing knowledge should be rejected entirely. It means there is the need to apply Western concepts carefully and critically by employing flexible, reflexive, humble, patient, and honest approaches in the research project. In the next section, I describe how I employed these decolonial approaches in my work.

The Research Process and Methods

Planning for Entry

As customs and practices in research demand, before one can effectively research at a particular field, the researcher must seek approval from a "gatekeeper," "a person or group that has the authority to approve research access" (Lindlof & Taylor, 2011, p. 98). For a researcher to know who the gatekeepers are in an institution requires that the researcher did initial inquiries. The researcher can read about the institution on the Internet before visiting the physical space. It also requires negotiations with gatekeepers and people of authority. In my case, I leaned on acquaintances for help. I spoke with my dad about my research and asked for possible ways that I could get access to the headquarters of the electoral commission. He, in turn, spoke with a friend at the electoral commission at a local level in the community I reside in. After my dad had spoken with his friend, he asked me to personally call him, introduce myself and the project and asked how he could be of help. I called and we spoke at length about my project. After our conversation, he asked me to send an email detailing my project and what my objectives were. I did that and after three days, I received a reply that he managed to contact someone who was willing to help. I contacted this person through phone calls and he promised to help me. Thus, after these initial contacts were made, and after I had received assurance that I could conduct my study, I travelled to Ghana at the end of November 2014 to collect the first round of data which formed the basis of my dissertation (Dorpenyo, 2016) and a follow-up study in the summer of 2017. On December 3, 2014,

I reported at the contact's office to introduce myself to him. He asked me to submit an introductory letter and my proposal for consideration. I contacted my supervisor through email, and she responded immediately with an introductory letter. On the December 8, I submitted my introductory letter, an abstract, and my proposal for consideration. Then, data gathering began immediately after I was given approval.

I must say that approval for this project did not come so easily. When my gatekeeper agreed that I could research at the Commission, little did he know that he had to seek approval from various heads of department. He also didn't know he had to follow some laid-down procedures at the Commission. So, when I visited the various departments, heads resisted the idea of giving me access to documents. They stated that they had not received any official letter from "above" indicating I will be collecting data. They needed to see that document before they could give me access to the information I needed for my project. I quickly went back to the gatekeeper and relayed this information to him. He responded promptly by calling the various heads to inform them about my project. After his phone calls, I had much success with data collection. I must also say that a part of me felt that the resistance was their skepticism to a Ph.D. student collecting data and working on a dissertation. Much of the resistance changed when I went back in 2017 to collect the second round of data. Heads became friendlier when I informed them that I was an assistant professor and that I have a Ph.D. in technical communication. Even though most of them did not know the field, because technical communication is an underdeveloped field in Ghana, they were comfortable providing information to Dr. Dorpenyo (as many of them called me).

Data were collected within 38 days at the headquarters of the Electoral Commission in Accra. The Electoral Commission, per the 1992 constitution of Ghana, is required to, among other things; "conduct and supervise all public elections and referenda," and to "educate the people on the electoral process and its purpose" (*1992 Constitution of the Republic of Ghana*, 1992). Thus, it has become the responsibility of the Commission to find ways to conduct "free, fair and transparent elections to the acceptance of all stakeholders" ("Mision, Vision and Functions,"). The Commission is constituted by seven members: a chairman, two deputy chairmen, and four other members all of whom are appointed by the president of Ghana. In the next paragraph, I discuss the various departments and how department heads helped in the data collection process.

Structurally, the commission is subdivided into seven different departments: the Elections Department, the Training Department, the Research, Monitoring, and Evaluation Department, the Finance and Administration division, the Human Resources and General Services Department, the Information Technology Department, and the Public Affairs Department and the Internal Audit. I place emphasis on these departments because I went to each unit to collect data. The heads of the various departments report directly to the Chairman of the Commission. Sections of the biometric adoption process were handled by the various departments. For instance, the Training Department was in charge of the training of Officials. The department was also involved in the writing of the instructional manual that accompanied the biometric technology; the Research Department analyzed and assessed the strength of the biometric system and presented information on findings to stakeholders; and the Public Affairs Department designed educational materials that educated the public on the relevance of the biometric systems and ways to go about the registration process. They also designed flyers that accommodated the technology to the general public.

Positionality

Decolonial methodologies emphasize the need for research to reflect on how they position themselves in regard to participants they work with. Smith (2012) stresses how necessary it is for the researcher to map his or her position and ideological perspectives on the map. Thus, positionality played a major role in my research design and data collection. As indicated by Agboka (2014) and Jones (2014) and as it is an important trait in ethnographic research (Smith, 2012) and feminist research methodologies (Haraway, 1988; Harding, 1991; Keller, 1996), the position and role that a researcher assumes are very critical to research practice and they are very paramount in conducting international technical communication research. I assumed two positions: an *insider* (Ghanaian by birth) and an *outsider* not only because I live, work, and pursued my graduate education outside Ghana, but because I was exposed to Western forms of thought through my education in Ghana. Thus, even though I have lived most of my life in Ghana, my identity as a Ghanaian has shifted between being a native Ghanaian at one point, and an outsider at another point. I am a subject of colonialism. In essence, I was and have always been bi-culturally situated. This is the more reason why it is not a given that my identity as a Ghanaian will lead to an accurate representation of local reality. Considering my Western frames of thought, I would say that it is a *non-sequitur*. Therefore, the fact that I am an insider (a Ghanaian) does not mean I can understand everything that happened at the Electoral Commission of Ghana. Such positioning would amount to arrogance on my part. As an insider, I had to constantly reflect on how I interacted with the participants I worked with. I could also not say that by assuming an outsider position, I would be able to make informed decisions about my participants. So, I constantly worked between the insider/outsider frameworks.

I refer to my bi-situatedness because it forms an important aspect of defining what the parameters are when conducting international technical communication research or when one employs a decolonial approach. For me, the researcher conducting research in a postcolonial context must assume a fluid position or identity. My situation is a tacit example. When I go to Ghana (the land of my birth), I am a national, a native who enjoys all the privileges that a Ghanaian must enjoy first before international. However, my identity changes to "international" or "immigrant" as soon as I land on American soil. Thus, my unstable, fluid identity shapes my research report. My take is that when my research is tagged as international research, then my identity as a Ghanaian is missing. But do I have to separate my identity as a Ghanaian from my work? When this happens, I reinforce the ideology that separates the researcher from the researched. I reinforce the dichotomy between subject and object. It is at the backdrop of this bi-positionality that I argue for a research methodology which will enable the researcher to discuss his/her fluid, unstable position. On the other hand, my work is international because I use Western frames to study a non-Western society. My identity is both articulated and disarticulated in both instances. Thus, bi-situatedness and positionality play major roles in international technical communication research.

Humility

Dr. Karla Kitalong of Michigan Technological University in the US usually says that "it is hard to learn anything when you are trying to look like the smartest person in the room." What this means to me is that one needs to be humble to learn or listen to ideas from people even if you know it all. You will never know what the other person has for you if you do not pay close attention to what they have or if you do not humble yourself. Humility applies to a lot of human relations, and it is a useful tool in conducting research. By "humility," I mean the ability to "relinquish" the researcher as an expert role in order to listen to participants' knowledge on issues related to one's research. Humility also means paying attention to your research data. On one of my visits to the Electoral Commission, my contact was at a meeting so I was asked to wait for him in one of the offices occupied by a woman and a man. Seeing how idle I was, the gentleman engaged me in a conversation. During the conversation, my research emerged and he asked more about what I was examining. He got very interested in it because he was one of those involved in the adoption process. Although the interview was not part of my research method, I let myself go and engaged in a question-and-answer conversation. During our conversation, he reiterated the very reasons why the machine was adopted. According to this gentleman, the adoption was necessitated by such malpractices as over voting, vote rigging, and impersonation. Even though these were not new to me, I humbly listened to him. He later told me that the adoption process could not solve most of the issues. Asked why, he said he could fault the process of registration. It was through him that I got to know that the registration for the 2012 elections was conducted in phases. The various ways that polling stations were earmarked for the registration were problematic, he stated. Citizens who were ignorant about the various phases lined up to register when they saw their friends and relatives at different polling stations lined up for the process. These citizens who registered, though it was not their turn to register, lined up again to register when the registration centers were moved closer to their polling stations. A situation he said could have been avoided if they really educated the citizenry on registration procedures.

He also revealed to me one of the reasons why the machines broke down. They identified after the elections that most Data Clerks did not follow the instructions that came with the biometric technology. Clerks would rather learn more about the system by doing than referring to the manual. In the process, most of the machines broke down.

Also, it was he who hinted that the research department, the public relations department and the training department played a major role in the adoption process. After our conversation, he led and introduced me to the heads of these departments. Further, in a casual conversation with one official of the Electoral Commission, he revealed that the biometric device was imported from one of the European countries even though he could not recall the name of the country. This official also revealed that most of the Data Clerks and Polling Station Officers were temporary workers hired by the Electoral Commission during elections. Most of the Officers were teachers who have IT knowledge. So, by humbly listening and unlearning what I have learned, and by decolonizing my mind, I got wind of what some of the causes of the breakdowns were and how some issues as impersonation and over voting reoccurred.

I must also add that humility means respecting the wishes of participants. For instance, when I went to visit one of the directors at the headquarters of the EC he declined to be part of the interview because he was scared I could use his words to hurt his reputation. Although I communicated to him strongly that everything will be confidential, he declined to be part of the study. I respected his wish and walked out of his office. Even though it was a painful experience, I accepted it in good faith. I demonstrate in the next paragraphs that it is not enough to be humble, a researcher must be patient and flexible.

Patience

Related to humility is patience, but mostly little is said about the relevance of it to conducting international technical communication research. A researcher conducting research in a postcolonial context is bound to face frustrations and the only way to overcome such vexations is when one exhibits patience. The researcher becomes more offended when they encounter such frustrations in their own countries. In instances where the researcher encounters frustrating moments, the researcher must learn to be calm and not become annoyed easily. I recall the many instances when I had to ask whether I would have been subjected to certain treatments if I were conducting research in a different country. I narrate some of these forms of frustrations and how with humility and patience a researcher could sail through a research process successfully. I stated earlier that my contact at the Electoral Commission asked me to submit my abstract, proposal, and introductory letter. My hope was that these documents were going to be attended to with alacrity. I was somehow disappointed when after three weeks of my stay in Ghana, the gatekeeper had not looked at those documents. Meaning that though I arrived early for data collection, I could not begin research immediately because the documents had not been approved. It took several calls and several visits to the commission.

It became more painful when after the documents were approved by the gatekeeper, various heads denied me access to their departments because they had not received any notice "from above" that I was conducting research at the headquarters. Immediately, I went back to the gatekeeper to discuss issues with him. Frustrated himself, and somewhat disappointed that he has delayed my research, he took it upon himself to distribute my documents to the various heads. I only had two weeks more left. Worst still, it took extra one week for heads to identify and locate most of the documents that I requested. I felt the pain and frustration. I wondered why I even decided to embark on the journey or why I decided to undertake such a project. But, I kept my cool, and constantly visited the Commission with the hope that some of the documents will be retrieved. I received most of my documents three days prior to my departure. It was a relief!

After I had finished collecting my data, I went back to my contact to thank him for his support. He apologized to me for the delay. According to him, he was new and did not know about some of the protocols at the Commission. He thanked me for not venting my spleen because he had witnessed the way some students abandoned their projects because they went through similar frustrations and burdensome bureaucratic procedures. To the best of his knowledge, I was patient with them. And I must say that this show of patience showed off as I got 80% of the documents I requested. I also became friends with most of the heads because I was calm with them and waited till they made the documents available. Some shared in my frustrations and encouraged me to contact them anytime I needed help.

Data analysis also requires the art of patience. For instance, it took me enough time to connect data to literature; I also had to struggle to make meaning of my data; and which analytical lens to use.

Reflexivity

One of the most cherished practices in conducting qualitative research is self-reflexivity (Tracy, 2010, p. 842), and it becomes an indispensable tool in international and intercultural technical communication research (Jones, 2014; Thatcher, 2010; Walton, 2014) or in decolonial methodology (Smith, 2012). According to Smith (2012), reflexivity is a critical tool which needs to be exercised constantly in the field. She indicates that "insider researchers have to have ways of thinking critically about their processes, their relationships and the quality and richness of their data and analysis. So to do outsiders" (p. 138). Self-reflexivity guides or pushes researchers to be honest with themselves, their research and their audiences. Self-reflexivity encourages "writers to be frank about their strengths and shortcomings" (Tracy, 2010, p. 842) because in reflexivity "you ask yourself the same questions that guide your analysis and other interpretations" (Goodall, 2000, p. 141). Reflexivity therefore becomes "a process of personally and academically reflecting on lived experiences in ways that reveal the deep connections between the writer and her/his subject" (Goodall, 2000, p. 137). It also becomes a tool that gives researchers the opportunity to reflect and understand the cultural constructions of their theories and methods… (Walton, 2014, p. 3). So how did my identity as a national (Ghanaian by birth) and an international student/faculty researcher shape the way I conducted the research? What are those things that are worth recording? What is the culture of the place and how does this culture help me to proceed with my research work? How did I even become interested in my project? How did I choose data collection method and analysis? I demonstrate how reflexivity worked vis-à-vis other research practices.

Flexible Design for Data Collection

Method

Data for analysis were collected on two levels between November 2014 and August 2017. In both cases, research was conducted after the general elections. The first form of data collection focused more on technical documents, that is technical documents that give "users control over technological systems, to empower" (Seigel, 2013, p. 34), and this was conducted through a dissertation project (Dorpenyo, 2016). Data were limited to technical documents such as instructional manuals, parliamentary reports, online videos, educational guides, and PowerPoint slides. In the dissertation project, my goal was to identify how the various genres were used as a means to communicate the biometric to the Ghanaian electoral and other stakeholders. In all, I collected 18 different textual materials. Since these documents were too much for my project, I decided on which documents were relevant and could help contribute to knowledge in the field. Eight criteria informed my decision. First, the document must meet my working definition of technical writing.

Second, the document should work to control/initiate an action. The document must be produced by or for the electoral commission and it should give an instruction about the biometric device or the biometric verification system. Fourth, the document should be targeted toward EC officials, Ghanaian who is of voting age, and data entry clerks. Also, the document should discuss issues pertaining to the 2012 elections. Last and most importantly, the document should make a case/argument for why it is necessary to adopt the technology. In this regard, the document "might be argument for constituting a technological system, maintaining it, disrupting it, or replacing it" (Seigel, 2013, p. 32). I settled on the user manual that accompanied the biometric device, parliamentary hansard (recordings of parliamentary debates), an online video of biometric voter registration education, and a suite of posters used to communicate issues about the biometric technology to the public.

After the analysis of these technical genres, I followed up with a second level of data collection which employed qualitative interview method. My goal here is to hear tales from biometric handlers and policymakers about the problems the country encountered during the integration process and how such problems were resolved in subsequent elections. In all, I interviewed 5 heads of department at the EC and 15 verification officers. Verification officers proved to be useful to this work because they used the biometric to verify voters on Election Day. Thus, they witnessed from below what issues hampered the use of the biometric, while the heads discussed policies initiated to help solve the problems. Heads provided a bottom-up perspective of the use and verification officers provided a down-up perspective of the verification. All subjects were selected using convenience sampling/snowballing approach (cite sources). For instance, my cousin who worked as an EC official during the 2012, 2016 elections helped me to contact some of his colleagues who were ready to participate in the research and a friend who was also close to some verification officers in a local polling station in the Western region of Ghana helped me to contact verification officers. It was not so difficult to interview the heads because I established some kind of a relationship with them when I visited Ghana in 2014 to collect the first level of data. I maintained contact through regular phone calls to them to ask how they were doing. I must say that some things had changed when I visited the electoral commission of Ghana in 2017. For instance, the Head of the Planning Department had gone on retirement and the Planning Department had been subsumed by the Electoral Services Department (it was called the Election Department when I went in 2012); Public Affairs Department had been renamed the Communications Department, and there was a new Chair of the Electoral Commission. More so, the new Chair and two of her deputies were accusing one another of fraudulent or corrupt means; the two deputies were asked to proceed on leave, while the president, the Chief Justice, and Parliament had been petitioned to impeach the Chair of the Commission (cite). On July 27, for instance, the Chair of the Commission was asked to appear before parliament (a meeting which I witnessed) to account for "the money collected from journalists for accreditation cards for the coverage of the 2016 elections, as well as that collected for the replacement of lost voter identity cards (IDs)." Regardless of these challenges, I managed to collect data as scheduled. I reveal these stories to speak to the fact that the interviews do not reflect the views of the entire commission but a carefully selected few.

Out of the 20 participants interviewed, only one was a woman and most of the verification officers were either first degree or master's degree holders. Others are also undergoing studies to obtain a first degree in different disciplines. Participants were not offered payments, and this was in accordance with conditions I stated in my IRB consent form which was approved by the IRB board at George Mason University (IRB #149249-I) in April 13, 2013. All interviews were face to face. Before each interview, I gave out the informed consent forms to interviewee's days before the actual interview process. I did this to ensure that interviewees understood the terms and the research project. Before each interview, I asked interviewees if they had questions about the research or anything in regard to the conduct of the research. After their questions had been answered, I asked them to sign the informed consent forms. Each participant signed two copies: one for them and one for me. Because the participants varied in age and status, I designed separate questions for them. For verification officers, I asked them questions about the problems they encountered when they used the biometric technology, what user problems were and their general impressions of the biometric technology. For management-level interviewees, I asked about the role they played in the adoption process, the problems the EC encountered integrating the biometric technology and how problems and how they managed to resolved problems such as breakdowns and rejections. In the next chapter, I will reveal some of their experiences and how such experiences inform discussions about localization and social justice, but in the next section, I discuss how I analyzed the interviews and documents I collected.

Flexible Analytic Tools

Because decolonial methodology employs flexible approaches, I was able to use two analytic tools to analyze data I gathered. I adopted and used Blake Scotts and rhetorical–cultural analytic tools to analyze the documents and grounded theory to interpret the interview data. The rhetorical–cultural approach provided a flexible lens for document analysis. As Seigel (2013) identifies, a technical communicator using rhetorical–cultural approach as an analytic lens "would read manuals [and genres] in order to begin to map how the user's identity has been articulated over time. Because they tend to focus on user tasks and actions, it is possible to investigate articulations by first paying attention to what user practices the manual facilitates" (p. 24, bracket mine). This way, the critic can *isolate, describe, classify, interpret,* and *evaluate* a phenomenon (Hart & Daughton, 2005, p. 25) under study. In this project, I perform a rhetorical–cultural analysis of technical documents to identify the role such documents played in maintaining assumptions about the biometric technology. In line with this goal, I critically work to identify implicit and explicit arguments about the necessity to adopt the biometric device and the consequence of adopting the biometric.

GROUNDED THEORY

I analyzed the data by loosely following a grounded theory approach (Charmaz, 2014; Strauss & Corbin, 1990). Grounded theory is a flexible method which enables researchers to pay attention to research data and use data as the prime basis for analysis, categorization, and theory construction. Charmaz (2014) defines grounded theory as a "systematic, yet flexible guidelines for collecting and analyzing qualitative data to construct theories from the data themselves" (p. 2). It is not formulaic, but it offers flexible heuristic guidelines for analyzing research data, and it also helps the researcher to see his/her data "in fresh eyes" and explore ideas about the data through analytic writing. Because it is an inductive process and because the goal is to construct theory from below, a grounded theory helped me to begin to study my interview data to make sense of how Ghanaian users encountered the biometric data. By focusing solely on carefully collected data, I was able to move from "particular to the more general;" "move up from the detailed descriptions to the more abstract, conceptual level" (Bryant & Charmaz, 2007, p. 15).

One can realize that grounded theory is an approach which supports decolonization projects. First, grounded theory expects researchers to be flexible and reflexive in data gathering and analysis. For instance, it is a normal practice for grounded theorists to "go back and forth between empirical data and emerging analysis" (Bryant & Charmaz, 2007, p. 1). I follow this approach because it advances the philosophy of decolonial methodology which emphasizes respect for participants, humility, flexibility, and reflexivity in data gathering and analysis. As a matter of fact Strauss and Corbin (1990) identify that grounded theory "help[s] the analyst to break through the biases and assumptions brought to, and that can develop during the research process" (p. 57). Because theory building is the core goal of grounded theory, researchers are able to pay particular attention to the data and the theories that emerge are from the data that were gathered not any imposed theory from outside. This ensures that participants' voices are captured in the process of knowledge construction. For instance, the three localization theories I advance in this book emerged from the data I gathered through interviews. By allowing data to speak to the theories I construct, I engage in an original project which seeks to truly empower the marginalized. More revealing is the fact that both decolonial methodologies and grounded theory seek to build knowledge from the ground up. Decolonial methodology, for instance, seeks to give a voice to indigenous languages, knowledges, and cultures that have been silenced or misrepresented, ridiculed, or condemned, while grounded theory acknowledges participants' experiences and constructs knowledge through participants' experiences and ways of knowing. I hold that it is only by carefully paying attention to and respecting the data generated through research that marginalize languages, knowledges, and cultures can be recovered. More so, paying attention to research data means acknowledging, local logics, histories, philosophies, economics, and politics in a colonized context.

Grounded theory helped me to construct knowledge from below through three stages: open/initial coding, axial coding, and selective coding. Coding is important to research because it opens up more avenues for inquiry. Coding helps the researcher to identify and "develop concepts and analytical insights through close examination of, and reflection on, field note (interview) data" (Emerson, Fretz, & Shaw, 2011,

p. 175). The goal of this process is to "produce coherent, focused analyses of aspects of the social life that have been observed and recorded, analysis that are comprehensible to readers who are not directly acquainted with the social world at issue" (cited in Emerson et al., 2011, p. 171). Following traditional coding method, I went through several processes to get my coding done (Emerson et al., 2011, p. 172; Rubin, 2005, p. 200; Saldana, 2009, pp. 73–84). For grounded theory, coding "is the central process by which theories are built from data" (Straus and Corbin, p. 57). Open coding (Straus and Corbin) helped me to break down, examine, compare, conceptualize, and categorize my interview data. I did this in order to (1) reduce the number of units in the data, (2) look for similarities and differences, and (3) ask questions about the phenomena reflected in my interview data. To accomplish this, I read and reread the transcribed interviews to highlight some major and interesting statements made by interviewees. I followed that by summarizing every response given to each question by the interviewees. After summarizing, I reread the transcripts again and used words and phrases to name, label, and categorize phenomena identified in data.

I followed the open coding process with axial coding, loosely based on Straus and Corbin (1990). Grounded theorists use axial coding when they want to put data "back together in new ways after open coding by making connections between categories" (Straus & Corbin, 1990, p. 96). This process helped me to specify a category (phenomenon) in relation to the conditions that gave rise to the phenomenon; the context in which the phenomenon is embedded; the actions/interactional strategies exhibited by actors to salvage a situation; intervening conditions and the consequences of the tactics/strategies deployed to manage the phenomenon. Figure 3.1 exemplifies the results of the axial coding.

In simple terms, the diagram means that the major PHENOMENON which characterized the use of the biometric technology was breakdowns; CAUSAL CONDITIONS include the breakdown of biometric technology and inefficient user manual; the CONTEXT indicates that biometric technology stopped working and it rejected people; the INTERVENING CONDITIONS are hot weather, rough/worn out fingers of voters, lack of proper biometric procedures, lack of technical expertise on the part of some EC officials; ACTIONS/STRATEGIES adopted to solve breakdowns include the use of tents to cool control weather, use of OMO or detergents to wash the hands of rejected voters, supply of back up devices in subsequent elections, intensified training of EC officials on how to

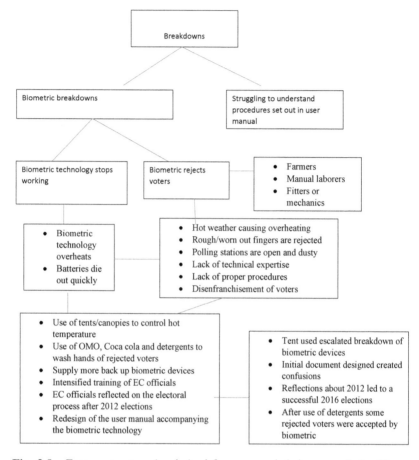

Fig. 3.1 Emergent categories derived from grounded theory analysis of interview data

handle biometric breakdowns, EC officials reflected on the entire electoral process, and the redesign of the user manual; the CONSEQUENCES of the actions taken are (1) tent used escalated the breakdowns, initial user manual redesigned for 2012 caused confusions, reflections by EC after 2012 elections led to improved 2016 elections.

After axial coding, I worked to select my core category, systematically related the core category to other categories, validated the relationships between them, and filled in categories that needed to be further refined and developed. This process is referred to as "selective coding." Straus and Corbin (1990) identify five stages of this process: (1) explicate a storyline; (2) relate subsidiary categories around the core categories by means of the paradigms developed in axial coding; (3) relate categories at the dimensional level; (4) validate those relationships against data; and (5) filling in categories that need further refinement (p. 118). So based on the axial coding above, I advance one single story about the biometric technology which was adopted by Ghana:

> the main story seems to be about the issues the EC encountered when it adopted the biometric technology for the 2012 and 2016 elections to enhance the electoral process of Ghana. My analysis reveals that during the electoral processes, especially the 2012 elections, the biometric devices broke down on several fronts: batteries died out and the biometric devices reject a group of people. Some issues accounted for these breakdowns: 1) weather was very hot for the devices and 2) some voters had rough fingers so they could not be identified by the biometric. When the biometric broke down, the EC of Ghana and voters devised several strategies to resolve the issues: they used tents/canopies to control the temperature, and they also used OMO, Coca-cola and other detergents to wash the hands of those rejected.

Based on this storyline, I identified one main idea: active management of breakdowns. This main idea helped me to come up with one core category: USER STRATEGIES in the midst of breakdowns. This core category will form the basis of two of the localization principles I advance in this book: user-heuristic localization (Chapter 6) and subversive localization (Chapter 7). User strategies, as I will advance in this book, do not only fit well into localization projects, but it also helps to advance the goals of decolonial projects of making visible local knowledge, culture, and histories of marginalized and or colonized societies. More so, this project pays attention to the self-determination of users in Ghana to enhance their electoral integrity, empowerment, decolonization, and the social justice implications of the biometric adopted. Grounded theory helped me to achieve these goals.

In Chapters 5 and 7, I perform rhetorical–cultural analysis of documents such as instruction manual, online video, parliamentary hansard, and a suite of posters to demonstrate how the rhetoric surrounding a technology can enable an articulation of broader issues.

REFERENCES

1992 Constitution of the Republic of Ghana. (1992).
Acharya, K. R. (2019). Usability for social justice: Exploring the implementation of localization usability in Global North technology in the context of a Global South's country. *Journal of Technical Writing and Communication, 49*(1), 6–32. https://doi.org/10.1177/0047281617735842.
Agboka, G. Y. (2012). Liberating intercultural technical communication from "large culture" ideologies: Constructing culture discursively. *Journal of Technical Writing and Communication, 42*(2), 159–181.
Agboka, G. Y. (2014). Decolonial methodologies: Social justice perspectives in intercultural technical communication research. *Journal of Technical Writing and Communication, 44*(3), 297–327.
Bryant, A., & Charmaz, K. (2007). *Introduction grounded theory research: Methods and practices* (A. Bryant & K. Charmaz, Eds., Paperback ed.). Los Angeles: Sage.
Charmaz, K. (2014). *Constructing grounded theory* (2nd ed.). London and Thousand Oaks, CA: Sage.
Comaroff, J., & Comaroff, J. (1991). Of revelation and revolution. In *Christianity, colonialism, and consciousness in South Africa* (Vol. 1). Chicago: University of Chicago Press.
Ding, H. (2014). *Rhetoric of a global epidemic: Transcultural communication about SARS*. Carbondale, IL: Southern Illinois University Press.
Ding, H., & Savage, G. (2013). Guest editors' introduction: New directions in intercultural professional communication. *Technical Communication Quarterly, 22*(1), 1–9.
Dorpenyo, I. K. (2016). *"Unblackboxing" technology through the rhetoric of technical communication: Biometric technology and Ghana's 2012 election*. Open Access Dissertation, Michigan Technological University.
Emerson, R. M., Fretz, R. I., & Shaw, L. L. (2011). *Writing ethnographic fieldnotes* (2nd ed.). Chicago: The University of Chicago Press.
Gonzales, L. (2018). *Sites of translation: What multilinguals can teach us about digital writing and rhetoric*. Ann Arbor: University of Michigan Press.
Gonzales, L., & Zantjer, R. (2015). Translation as a user-localization practice. *Technical Communication, 62*(4), 271–284.
Goodall, H. L. J. (2000). *Writing the new ethnography*. New York: Rowman and Littlefield.
Haraway, D. (1988). Situated knowledges: The science question in feminism and the privilege of partial perspective. *Feminist Studies, 14*(3), 575–599.
Harding, S. G. (1991). *Whose science? Whose knowledge? Thinking from women's lives*. Ithaca, NY: Cornell University Press.

Hart, R. P., & Daughton, S. M. (2005). *Modern rhetorical criticism* (3rd ed.). New York, USA: Pearson Education.

Johnson, R. (1998). *User centered technology: A rhetorical theory for computers and other mundane artifacts*. Albany: State University of New York.

Jones, N. N. (2014). Methods and meanings: Reflections on reflexivity and flexibility in an intercultural ethnographic study of an activist organization. *Rhetoric, Professional Communication, and Globalization, 5*(1), 14–43.

Keller, E. F. (1996). Feminism and science. In E. F. Keller & H. E. Longino (Eds.), *Feminism and science* (pp. 28–39). New York: Oxford University Press.

Lindlof, T. R., & Taylor, B. C. (2011). *Qualitative communication research methods* (3rd ed.). Los Angeles: Sage.

Lovitt, C. R., & Goswami, D. (1999). *Exploring the rhetoric of international professional communication: An agenda for teachers and researchers*. Amityville, NY: Baywood Publishing Company.

Rubin, H. J., & Rubin, I. S. (2005). *Qualitative interviewing: The art of hearing data* (2nd ed.). Thousand Oaks, CA: Sage.

Saldana, J. (2009). *The coding manual for qualitative researchers*. Thousand Oaks, CA: Sage.

Seigel, M. (2013). *The rhetoric of pregnancy*. Chicago: University of Chicago Press.

Smith, B. (2012). *Reading and writing in the global workplace: Gender, literacy, and outsourcing in Ghana*. Lanham: Lexington Books.

Strauss, A. L., & Corbin, J. M. (1990). *Basics of qualitative research: Grounded theory procedures and techniques*. Newbury Park, CA: Sage.

Sullivan, P., & Porter, J. E. (1997). *Opening spaces: Writing technologies and critical research practices*. Greenwich: Greenwood Publishing Group.

Sun, H. (2004). *Expanding the scope of localization: A cultural usability perspective on mobile text messaging use in American and Chinese contexts*. Troy, NY: Rensselaer Polytechnic Institute.

Sun, H. (2006). The triumph of users: Achieving cultural usability goals with user localization. *Technical Communication Quarterly, 15*(4), 457–481. https://doi.org/10.1207/s15427625tcq1504_3.

Sun, H. (2009). Toward a rhetoric of locale: Localizing mobile messaging technology into everyday life. *Journal of Technical Writing and Communication, 39*(3), 245–261. https://doi.org/10.2190/TW.39.3.c.

Sun, H. (2012). *Cross-cultural technology design: Creating culture-sensitive technology for local users*. New York: Oxford University Press.

Thatcher, B. (2010). Editor introduction: Eight needed developments and eight critical contexts for global inquiry. *Rhetoric, Professional Communication, and Globalization, 1*(1), 1–34.

Thatcher, B. (2011). *Intercultural rhetoric and professional communication: Technological advances and organizational behavior.* Hershey, PA: IGI Global.

Thrush, E. A. (1993). Bridging the gaps: Technical communication in an international and multicultural society. *Technical Communication Quarterly, 2*(3), 271–283.

Tracy, S. J. (2010). Qualitative quality: Eight "big-tent" criteria for excellent qualitative research. *Qualitative Inquiry, 16*(10), 837–851.

Walton, R. (2014). Editor's introduction to the special edition on methodology. *Rhetoric, Professional Communication, and Globalization, 5*(1), 1–13.

Young, I. M. (1990). *Justice and the politics of difference.* Princeton, NJ: Princeton University Press.

CHAPTER 4

Stories of Users' Experiences

I must be clear about my use of the term "users." I identify two levels of users in regard to the biometric use in Ghana. The first group of users belongs to the category of voters. A voter can be considered a user of the biometric technology since that person must be verified and authenticated before he/she can vote. In this regard, voters are partial or secondary users of the biometric. The second group of users belongs to the decision-making body of the Electoral Commission (The EC Chair, Heads of Department, and District Electoral Officers) and operational staff, notably verification and registration officers. The EC officials are the primary users since they handle the biometric device and make sure that users are able to use the system to vote. I mainly interview this second group of users. Their stories form the basis of my discussion in current and subsequent chapters.

Kofi Manu's Story: Desiring to Be Part of the Biometric Design Process

if we want to bring in some equipment for the improvement of our electoral system…definitely if the commission is actually part of it, we will ask: will this equipment withstand heat, cold or whatever it is

Profile

Kofi Manu is a high-ranking member of the Electoral Commission of Ghana. He is one of the few people who rose up through the ranks of the EC to the top. He started as the District Election officer of the EC in a small city called Windy bay. After serving in the position for a while, he became the Deputy Regional Director and later Regional Director. After spending two years as the head in this region, he was transferred to the region of gold where he worked as the Regional Director for two years and later transferred to the headquarters. At the headquarters of the EC, he heads the department which is considered to be the "engine" or the "heart of the Commission." His department:

- Draws and formulates policies for the operation of activities of the commission.
- Draws program of activities that guide electoral exercises—registration of voters, supervision of internal elections of political parties, supervision of the election of regional representatives of the council of state and all statutory elections in Ghana, including presidential and parliamentary elections, and the conduct of referenda.
- Works with the finance department to come up with a budget for general elections.
- Distributes and retrieves election materials
  - Prints ballot papers and distribute to constituencies.
- Gazettes elections that are held, that is, publish full results of elections on their website, journals, and the media.
- Registers political parties.
- Gives accreditation cards to observers.

I met Kofi Manu in November 2014 when I went to Ghana to gather the first level of data for my dissertation. Because of my prior engagement with Mr. Manu and because I maintained my contact with him after my first level of data collection, it was easy to arrange an interview meeting with him. Mr. Manu presents an interesting perspective about the biometric technology and the entire electoral system. My interview with him underscored one fundamental issue: The EC of Ghana suffers a lot of challenges which hinder electoral integrity. He revealed some of the challenges of the EC as: minors and foreigners voting during elections, the disruption of the electoral process by thugs/macho men, and perceptions (public perceptions, voters' perceptions, political parties' perceptions) about the electoral process.

Even though the electoral system suffers such challenges, he revealed that the EC has taken steps to manage some of the challenges. He revealed how the EC moved from using opaque ballot boxes to transparent ones because of the mass perception that voters carried and dropped foreign materials into ballot boxes; the EC moved from using thumbprint voter cards to photo IDs; institution of such structures as the Inter Party Advisory Committee (IPAC), District Inter Party Advisory Committee (DIPAC), and Regional Inter Party Advisory Committee (RIPAC). This was to ensure that stakeholders played a major role in the electoral decision-making process.

Mr. Manu believes that the biometric was able to solve two major challenges that most people were worried about: double registration and multiple voting. But he was quick to add the challenges the new technology introduced:

- Because of faulty printers and computers, when you key in specific data, the machines froze, and this caused some election officials to reregister some voters.
- The machines could not withstand the hot weather in the country, so it performed poorly when it was used under the heat of the sun.
- Users failed to obey instructions, and this led to breakdowns of the biometric.
- The batteries also died out frequently, and it took a lot of time to reboot whenever batteries were changed.
- The biometric also rejected people who had rough fingers.

Apart from these technical challenges, Mr. Manu also revealed they had to face local cultural myths about biometric use. They identified that some people failed to register and vote because of the perception that:

- If a pregnant woman goes through the biometric registration process, her pregnancy will be terminated or will experience complications.
- If you are a man, you will become impotent.

When these challenges occurred on Election Day, the EC adopted ad hoc strategies to salvage the situation and to ensure that elections go on unhampered. One of such strategies was the use of canopies/tents to control the hot temperature. Mr. Manu also revealed that after the 2012 elections, these other steps were taken:

- EC officials held meetings with regional directors to receive reports on incidents that happened during the election.
- There were calls from civil societies and EC officials met to ponder over recommendations offered by the Supreme Court after the historic election petition.
- The EC officials pondered over these issues:
  – Why did the biometric broke down?
  – Why did some EC officials fail to sign pink sheets?
  – Why did over voting occur?

After the EC had pondered over these issues, they identified, for instance, that biometric technology broke down because of these reasons:

- Some EC officials were not able to handle device properly.
- Some officials did not follow the appropriate operational procedures.
- The weather was not conducive for the device.

For these reasons, the EC officials resolved to (1) increase their training days to ensure that users were conversant and comfortable using the design; and (2) supply 2 BVDs to each of the polling stations in Ghana; ensure that well-qualified individuals were employed to handle the biometric devices. The consequence of these measures taken was a much improved electoral process in 2016. Reports indicate that 2016 recorded less breakdowns and officials showed familiarity with the device and the procedures.

Mr. Gadugah's Story: Learning from Previous Challenges

So learning from the challenges of 2012, when we went into 2015 there was an improvement

Profile

My Alfred Gadugah is a middle-aged man and has worked with the Electoral Commission of Ghana for twenty-one (21) years. He started as the District Officer then he became the Senior Electoral Officer, then he became the Principal Electoral Officer, Assistant Director of his department in May 2016, and presently serving as the Head of Department. As the HOD, he oversees the general administration of the department, coordinates with the human resources, and manages internal communication and media relations.

Mr. Gadugah believes that the EC has taken initiatives which have improved operations and the management of elections in Ghana. The major issue hampering credible elections is the enormous number of illiterate citizens who only can't read, but also are poorly informed about the basics of making choices. In addition to the illiteracy rate, he also believes that the media, political parties, and stakeholders are to be blamed for Ghana's electoral woes. There are also perceptions that the EC is corrupt and rigs elections to favor political parties or candidates. But he acknowledged that the EC has built a robust system which ensures that no individual could rig elections in Ghana.

Mr. Gadugah believes that the biometric has been able to resolve two major issues in the electoral process: impersonation and multiple voting. Beyond these two issues, the biometric was ineffectual in other instances, for instance, it cannot detect minors and foreigners who fraudulently register to vote. In spite of these downsides to the biometric use, Mr. Gadugah also identified that when the technology was adopted, the EC had to overcome one major problem: voters' fear that the biometric spreads HIV.

He acknowledges that the biometric technology broke down during elections due to some reasons:

- Inadequate education
- Limited piloting of the device
- Inefficiency of the system.

Mr. Gadugah revealed that when the biometric was first implemented, the "breakdown rate was just too high" and people couldn't use it. The consequence was that people got disenfranchised. He emphasized that biometric kit was not robust enough so the printers were just breaking down because they could not take the pressure; operational instructions were not clear enough and caused a lot of confusion; technicians also blamed it on humidity/high temperature.

After the elections, the EC sat down to reflect on the causes of the breakdowns during the elections and took these measures to prevent subsequent occurrences:

- They developed an intensive program to educate the populace of the biometric systems and its role in the electoral process of Ghana.
- Proper training of the EC officials handling the biometric devices.
- Learning from challenges.
- Decided to provide every polling station 2 BVD in the 2016 election.

According to Gadugah, learning from previous challenges informed:

- Training
- Deployment of machines
- Quantity of BVDs used in subsequent elections.

He believes that breakdowns are normal circumstances that characterize technology transfer and adoption. To him, "over the years, we adapt it as well as we use it and see the challenge, then we improve upon it to fit our environment." He funny it funny that biometric breakdowns will be attributed to humidity/high temperature because most laptops used in Ghana are designed outside the country but most educated Ghanaians use laptops anyway. So the breakdowns cannot be attributed to high temperatures, the biometric was just not "robust enough." He found it funny that when during the initial training sessions, trainers kept insisting that the biometric users should not go and put the machine in direct sunshine. This order was funny because "Ghana votes in the open. We have our polling stations on corridors, under canopies and when you keep telling us that we shouldn't put it in direct sunshine, do you think somebody in their right frame of mind will carry this thing and go and sit in the direct sunshine?"

He also revealed that the canopies provided to control the warm temperature even escalated the breakdowns because "if you ever sat under these canopies, the heat that the canopies radiate is even more than going directly under the sunshine." Barring these inefficiencies, he was of the view that if the designers had started their biometric production from "the bottom knowing the peculiarities of our environment then that will inform the way they [biometric] function." He continued that

> there is the need for suppliers and the vendors to take note because we don't vote in air-conditioned environment, okay, and in December, temperatures are very high, dusty environment...We vote in December, harmattan and when you ever go to Western or Northern region, you see the dust coloring the whole skyline. The whole skyline becomes yellow, brownish yellow because of the dust so in some of the areas you can't avoid the dust in December when you are voting.

When I asked whether the biometric rejected a specific group of people, he responded that there were speculations about the demography of people who were rejected but in the district he was assigned to supervise, they identified the biometric rejected cocoa farmers. This was the case because most of the cocoa farmers had lost most of their fingerprints owing to the nature of their job. On how those rejected had the opportunity to vote, he stated that "we heard that when people washed their hands with Coca-Cola" and so they also started that process and to their amazement the machine "started picking up."

Mr. Kofi Ansah's Story: Desiring to Know How Technology Was Built

Profile

Mr. Kofi Ansah has worked with the EC of Ghana for 20 years. He is the District Electoral Officer and his roles are to:

- Organize and supervise elections
- Oversee voter registration
- Recruit officials involved in the electoral process
- Organize training sessions for employees before elections
- Make sure election materials are deployed to all polling stations in his district
- Oversee two constituencies.

Like Manu and Gadugah, Ansah believes that one of the major issues that undermine election integrity in Ghana is that suspicions political parties harbor against one each other. Each party believes that "one way or the other somebody may try to outwit the other." He quells such suspicions by indicating that there is no need for such attitudes since the EC has in place mechanisms that ensure transparency and honesty.

Asked to discuss how useful the biometric has been to Ghanaians, Ansah indicated that the EC adopted the biometric for two reasons:

- People were complaining of over voting, or that people impersonated others so in order to check these ills, the biometric was brought in.
- The EC also wanted to know the total number of registered voters. This ensures that people do not register/vote twice during an election.

Like other interviewees, he was optimistic that the biometric has been able to resolve these two issues. In spite of his optimism, he was also quick to discuss the challenges the country faced when the biometric was introduced: "in 2012 what we experienced was that the BVD, well we did not know how the system was built, but many people could not vote. You place your thumb on and it said you cannot be verified." He also recalled that some of the individuals who suffered rejections were "farmers, fishermen, and the fitters (mechanics)." He also revealed that some individuals had perceptions about the biometric technology as they believed that technology has the ability to detect or identify those ladies who are pregnant.

Interestingly, Ansah indicated that when the biometric rejected people, they tried to encourage them to clean their hands with Coca-Cola, Fanta, tomatoes, etc., but those proposed measures did not work. Individuals who tried using these unconventional means proposed still suffered rejections when they tried voting for the second time. This revelation contravenes stories told by Gadugah and Manu.

Francis Yankson's Story: The Nature of the Heat Was Too Much

Profile

Francis Yankson graduated with a Bachelor of Science degree in 2009 in Computer Engineering from the Kwame Nkrumah University of Science and Technology. He teaches ICT in a high school in Mr. Ansah's district. When he heard that the EC in his district needed more hands to help in the elections, he applied for a position and he got a spot. He was first assigned as the education officer on special duty and later reassigned as the verification officer on special duty. As a verification officer on duty, he had to exhibit knowledge of the biometric device since he was the first point of call when the device encountered issues. He was also the intermediary between poll workers and district election officers. To demonstrate his knowledge of the biometric device, he took me through how the biometric functions:

> you know the device has the scanning aspect where it will scan the barcode of the voter and so when it scans the barcode of the voter the biometric features of the voter will appear on the display we call it the display so the verification officer will then look at the person standing in the face and look at the picture on the device because before then during the registration the data of all the voters have been put on it already so when you put your finger on it and then the details are there per your thumbprint it verifies you

Yankson identified that the biometric technology was helpful to the electoral process of the country. According to him, the biometric came to take away issues such as multiple voting, multiple registration, and impersonation. He stated that prior to the adoption of the biometric, people could vote 2, 3, 4 times and still have the chance to vote but with the biometric, such illegality will be detected. Although he touted the importance of the technology, he also identified some challenges: "the main problem was the rejection. We also realized that when the sun was scorching then the device heats up and then freezes up" to the point that "it wouldn't respond to your instructions." He adds further that even though it is not a good news for the technology to breakdown when it was needed most, one cannot do too much about it because we are looking at a situation where the "nature of the heat is too much. When you put it in a very hot environment then per the level of how hot the place is then you will have such a problem." He reveals that in situations where they put the technology under a shady area, such problems did not occur. He also reveals one important issue which enhances our knowledge of localization: "technology comes with money so perhaps the decision to use this one was looking at financial situations because other technologies may have been expensive to purchase because the idea was that we could help control these dusty and hot conditions."

He did not only discuss weather-related challenges, but he also identified that the biometric rejected some individuals. When I probed further, he revealed that those who suffered rejection were those who "had dirty hands. Others too per the work they do—you know we have people who have been using their hands like the carpenters, seamstress, etc. With the seamstress they handle the scissors so they have problems with their furrows and for the hair dressers, the chemical that they handle do interfere with the device and the mechanics, because of the dirty oil and those things they normally handle."

Aside from these technological challenges, one other challenge he hinted at was the inefficiencies exhibited by poll workers. According to him, some of his colleagues had little knowledge on how the biometric functioned so they committed "petty petty mistakes." In order to make the biometric function properly to meet the various contextual exigencies, he believes it will augur well to include users in the design of the technology: "it is very good when you involve users in the development of the device, it's very very important."

Reflections on Biometric Use in Elections

So what do these stories tell us about users and biometric technology design? From the point of localization, these interview stories indicate that users have knowledge of use and they adopt technologies that could solve their local needs. As I indicated in Chapter 1, biometric technology, historically, was not designed to be used as an election technology. Its basic aim was to provide secure access or a means of identification. For instance, Britain developed some kind of biometrics so they could keep track of their colonial subjects in India. Moreover, it is normal to go through biometric scans at airports, borders, or restricted buildings.

In recent years, election management bodies have realized the potential of the biometric and have moved to adopt this technology to enhance their electoral integrity. In this regard, the use of biometric ushers societies into "a new era of surveillance in which the body is both the target and the instrument of control" (birth of biometric, p. 9). The core issue is that biometric has affordance which supports its use in election. Smartmatic, a biometric technology production company, argues that the biometric they produce "gives authorities all the hardware, software and services that they need to successfully manage each phase of the election process. We focus in designing technology that guarantees more efficient and transparent elections;" and GenKey argues that its technology is designed "to securely capture and verify identity information, using multiple biometric modalities and biographic data." Therefore, biometric has features which ensure transparency, efficiency, and integrity in elections, as well as in other areas which require secure access.

Therefore, by adopting the biometric, Ghana links the instrumental features of the biometric with the local electoral needs. The interviews reveal that Ghana adopted the biometric for a specific reason: to help resolve some electoral malpractices. Though interviewees indicated that the biometric could not eradicate social practices such as minors and foreigners voting, they were sure it was going to resolve impersonation and double registration. This is a testament that biometric use, just as any technology, is situated in a complex context, and the meanings users attach to those technologies are very necessary to the localization project.

The adoption also reveals that users are able to change the purpose integrated into the design of technology. Although the original purpose of the biometric is to secure borders and posts, users are able to use the affordances to achieve a different goal, thereby giving a different purpose to the technology. Ghana needed a system which could breathe transparency, accuracy, and efficiency into the electoral system and they found that in the biometric technology.

Although there were setbacks in the process of integrating the biometric technology into Ghana's electoral process, users' ability to repurpose and reconfigure the use of the biometric is relevant and must be of important concern for localization scholars. It is for this reason that I propose and advance subversive localization. The ability of users to subvert technology and reconfigure it is a worthy addition to conversations about localization. It is this process which reveals users as knowledgeable, creative, and innovative agents of technology design. The level of subversion must not be as grandiose as redesigning the biometric but can be as small as changing the user manual accompanying a technology or using the technology for purposes either than those configured by the developers.

The interviews also reveal how biometric was influenced by physical, use, and cultural contexts. For instance, interviewees indicated how the biometric could not withstand the harsh weather conditions in Ghana; how the voting officials struggled to understand the instructions accompanying the biometric and even understand the various functional elements of the biometrics. These are some factors which hampered the use of the technology during the elections. In the chapters that follow, I discuss how the technical documents surrounding the biometric technology and the interview data help me to discuss the various localization processes I advance in this book.

CHAPTER 5

Linguistic Localization: Constructing Local/Global Knowledge of Biometric Technology

I indicated in Chapter 1 that technical communication scholars have criticized traditional approaches to localization for being too narrow. One aspect of the localization process which has seen much criticism is translation, that is, "the process of converting written text or spoken words to another language" (Esselink, 2000, p. 4). This is because this process only involves changing "graphical user interface (GUI) components of a software application, such as dialog boxes, menus, and error status messages" (Esselink, 2000, p. 57) to meet the needs of targeted audiences. TPC scholars such as Agboka (2013) have argued that a narrow focus on linguistic features alone hurt localization processes because such an approach has the tendency of overshadowing broader issues that influence localization processes. Meaning, if localization focuses narrowly on language, issues like social justice, local economy, local politics, local knowledge systems, and legal concerns are hardly discussed. But he is quick to point out the important role language plays in conversations about localization. I must say that even though I agree that localization should not be solely focused on language, the domain of language should still play an important role in localization as it affords researchers and localization experts the opportunity to understand how users relate with technology. Gonzales and Zantjer (2015) have come to the defense of those who lean toward localization translation by arguing that contrary to the perceived notion that translation is about replacing "one word in one language with a word in another language," translation "exemplifies the complex negotiation of history, culture, and language..." (p. 280). This means that a focus on language can reveal knowledge systems in a locale. They show the way by examining the rhetorical strategies multilinguals adopt when they are translating materials from one culture to the other.

I do agree with Gonzales and Zantjer when they say that translation reveals local knowledge systems. I think the main issue with those who criticize a language focus has been that such scholarship follows a top-down approach, that is, these scholars usually tend to examine documents developed by designers who, in most cases, belong to cultures that are vastly different from the culture of users. For example, Agboka (2013) studies documentation, accompanying sexuopharmaceuticals which flooded Ghana's market. Interestingly, the documents were designed in China and transported to the Ghanaian market. In addition to the fact that the documents studied are designed in designer's culture, many scholars in our field do not study the various ways non-Western users represent the technologies they adopt. A much closer field to technical communication, rhetoric of technology, which studies how technologies are represented even do so by examining the language of designers and not that of users. Miller (1994), for example, examines the language used to represent the Japanese fifth-generation computer systems they developed. To this end, technical communicators and rhetoric of technology scholars are complicit in colonizing users' way of knowing.

What if we take a bottom-up approach to the study of documents? By this I mean, what happens and how will we perceive the role of language differently if we paid attention to documents designed by local users themselves? I argue that such an approach will reveal broader historical, social, economic, and political issues in a local. Actually, a bottom-up approach is what a decolonial approach to localization favors and, in this chapter, I indicate that when attention is paid to documents designed by users, we are opened up to broader issues such as local logics, local knowledge systems, and philosophies. We will realize that localization based on the study of language provides a barrage of information about users and their rhetorical strategies.

Therefore, in this chapter, I focus on technical documents, broadly defined to include multimedia texts, designed by EC officials to communicate the role of the biometric technology in Ghana's electoral process. As is revealed through the conversations in the previous chapter, the EC and major political stakeholders had hopes about the biometric technology which was adopted for the elections. This trust in the biometric technology is manifested through the various communication modes used. When the biometric technology was adopted by Ghana, the Electoral Commission and stakeholders designed technical documents—parliamentary hansard, user manuals, suite of posters, and an online educational video to educate and persuade the electorates into accepting the new system. To an extent, EC officials and stakeholders designed rhetoric to accommodate the biometric technology to Ghanaian voters by emphasizing the various features of the technology. By this I mean, the documents presented a picture of "how the technology 'should' work" (Kimball, 2017, p. 3).

A typical example of the use of positive words to describe the biometric is captured in the two-minute online video: "Biometric System Voter Registration Process" developed by SuperLock Technologies Limited (STL), the organization that provided the biometric to the Electoral Commission. While on the surface, the goal of the video was to educate the electorate on the electoral process and how important the biometric was to the success of the elections, on a deeper and theoretical level, the rhetoric surrounding the biometric shed light on the ideological construction as well as how local culture shapes the communication of a technology. As Miller (1998) points out in her analysis of the historical development of technology in American culture, different phenomena constitute the rhetoric of technology: "rhetoric about technology: often captures public representations or debates in public policy forums; rhetoric within technology: private proprietary discourse by which technological work 'gets done'; and rhetoric from technologies: the ways in which values and thought patterns developed by technological work extend to and prevail other cultural arenas" (p. 307). Two of the three characteristics described by Carolyn Miller—rhetoric about technology and rhetoric from technology—operated in the construction of the biometric technology and in how it was circulated and accepted in Ghana. Let us look at how the biometric was represented in four main documents: online video, parliamentary hansard, a user manual, and a suite of posters.

Online Video: Accommodating Biometric to Users

The video opens with some electorates lined up under trees and three others seated waiting to go through the registration process, while two registration officers attend to two voters. Among those standing in the queue were a man and a woman who seemed confused about something and needed clarifications. Then, a fairly old man (who goes by the name Kofi Ghana) who is dressed in a kente (a traditional Ghanaian-made cloth which is usually worn by chiefs and the wealthy in the society), a native sandal and a crown (which symbolize kingship) walks to the two confused voters standing in the queue. On seeing the old man, the confused man reached out to Kofi Ghana and asked what the biometric registration was about. Below is a transcript of the conversation that took place between them:

Man: Kofi Ghana, what is this biometric registration all about?
Kofi Ghana: [laughs and says] oh it is very simple. Biometric registration is simply the collection of information that is only connected to you and no one else like your fingerprints and your photographs so that when we say it is you we know it is you.
Woman: Kofi so what do I need to register?
Kofi Ghana: Registration? Simple! First you must be a citizen of Ghana, eighteen years or older and you must present some form of identification like your old voters ID, National ID, passport, National health insurance card, driver's license or two guarantors to vouch for you.
Man: so how does the registration process work?
Kofi Ghana: oh...very simple. The registration process is in three stages. Stage 1 is to complete the form with all your personal details like your name and age and so on...; stage 2 is where the registration officer enters all your information onto a laptop with your fingerprint capture using a fingerprint scanner, your photograph using a webcam and then a printer prints out all the details for your voters card; stage 3 is the final stage where your voters ID is prepared for you to collect, Simple!
Woman & man: wow, so is really that simple!
Kofi Ghana: No biometric verification No vote. One man one vote that counts.

The video is short, but it reiterates some assumptions about the biometric technology: It captures accurate results; it collects and stores data that are unique to one individual; every individual goes through three stages that ensure that data are accurately captured. Therefore, the biometric cannot fail because it possesses in-built mechanisms which ensure success. The next document is the parliamentary hansard.

Parliamentary Hansard: Deliberating on the Biometric Law

The parliamentary hansard is "the official nearly verbatim report of proceedings of the House. It is a repository and reflection of the legislative activities of Parliament."[1] The parliament of Ghana forms the legislative arm of government. The 1992 constitution of Ghana enshrines the Parliament of Ghana with powers to "make laws" which "shall be exercised by bills passed by Parliament and assented to by the President" (p. 72). The legal mandate of Parliament was expressed fully in the hansard as several of the Parliamentarians who contributed to deliberations about the Constitutional Instrument emphasized they were giving "a legal effect to the introduction of the biometric registration and verification of registered voters;" or that they were giving "legal teeth to the biometric." Hence, the hansard, the official records of parliament, recorded deliberations about the role of the biometric technology in Ghana's elections. The hansard I analyze is a 30-page document which contains proceedings of September 25, 2012. Titled "Parliamentary Debates: Official Report Emergency Meeting," this document captures several motions tabled on September 25, but of importance to this project is the motion: "Public Elections Regulations, 2012 (C.I. 75)." The motion was a report submitted to the House of Parliament by the Subsidiary Legislation Committee on the Public Elections Regulations, 2012 (C.I. 75) by Honorable John Tia Akologo on behalf of the Chairman of the Subsidiary Committee. The focus was to enact a law that will enable the EC to "make new Regulations to replace the existing C.I. 15 in order to...give legal effect to the introduction of the biometric registration and verification of registered voters..." (p. 785). Through deliberations, we identify that the EC of Ghana, since 1996, has embarked on reforms that will address issues that undermine the integrity of elections in Ghana. One such reform was the adoption of the biometric technology.

Because of its deliberative nature, the primary stasis question of the parliamentary debate about the biometric technology was procedural: How can we give "legal teeth" to the biometric technology? This was necessary because the existing Constitutional Instrument (C.I 15) did not make any room for the biometric to operate. C.I 15 was the Public Elections Regulations that was used to regulate Public Elections since 1996 (periods before the adoption of the biometric technology). Thus, with the adoption of the biometric, it was necessary to replace it with another Constitutional Instrument (C.I. 75), so that the biometric can be operational. During this procedural moment, several of the members of parliament represented the biometric as the kairotic intervention to our risky electoral process. The biometric, to an extent, becomes a "desperately needed tool for intervention and control of an epidemic that was out of control" (Scott, 2003, p. 208): electoral malpractices.

Basically, parliamentary proceedings defined the electoral process as at risk or Parliamentarians reinforced the perception that the general election was in crisis and that it needed to be saved through biometric verification. The biometric, it was hoped, would provide a moment of change in the electoral process of Ghana. For the biometric technology to be effective, it needed a legal backing and Parliament, the constitutional arm of government enshrined to enact laws made this possible. Through Parliamentary deliberations, we encounter how a technological mindset affected the laws that were enacted to see to the successful use of the biometric technology. We realize that sometimes technological mindset encourages "somnambulistic" laws that stymie critical questioning of the technologies we adopt and use. Similar forms of representation are visible in the suite of posters.

Suite of Posters: Educating the Populace on Biometric Registration

In this section, I look at how the biometric was represented in mundane documents like a poster. This document consists of nine posters bound together. Its design makes it suitable to be hung on walls like one would hang a calendar on a wall. The purpose of this document is to educate the populace on the biometric voting system. It is targeted toward a heterogeneous audience considering that the Ghanaian electorate comprises of either literates, those who can read and write English, semi-literates, or illiterates, those who have not had any form of formal education. In a way, this document reinforces arguments made in the online video. The first page, Fig. 5.1, for instance, defines what biometric registration is, what the biometric technology is and the purpose of the biometric in Ghana's elections. In this document, biometric registration is defined as "the capturing of some biometric features of voters during the registration process." Biometric technology is defined as "the use of computers

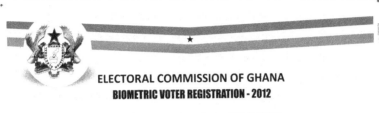

Fig. 5.1 Educational material defining biometric technology

to identify persons by means of their unique physical or behavioural features; such as face, iris, palm, fingerprints, voice…;" we are told that in Ghana's case features will be limited to fingerprints and photographs of the applicant. Also, the first page tells us the purpose of the biometric: "to prevent incidence of multiple registrations" and "to serve as a foundation for establishing a continuous voter registration."

The second page of the document establishes the similarities and differences between the old system and the new system of registration (Fig. 5.2).

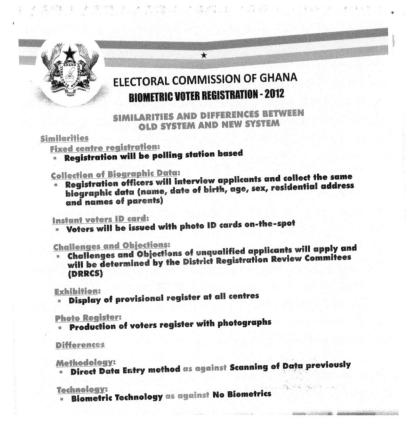

Fig. 5.2 Similarities between old and new registration systems

Some other important items stated in this poster include: qualification for registration; evidence of eligibility; registration process; process of challenging an applicant's registration; general advice; importance of voters' register. Worthy of note are statements such as the biometric technology "provides transparency in the verification of Election Results" and the general advice to electorates to desist from "multiple registration, since the Biometric voter registration system will expose" them. Below are samples of the posters (Fig. 5.3).

Fig. 5.3 General advice to voters

THE USER MANUAL: FOLLOW THE INSTRUCTIONS
TO OPERATE BIOMETRIC SMOOTHLY

The manual is a 31-page booklet produced by the Training Department of the Electoral Commission and STL, the providers of the biometric device. The manual serves as a training guide and a reference tool for operational staff, notably key trainers and polling staff who will handle the biometric device at the various polling stations. It mostly contains the procedures and legal precedents that articulate the operation of the biometric device. Words and phrases used are meant to identify the different aspects of the process in using the biometric device and what polling agents should do when they encounter such issues as the inability of biometrics to identify voters. Later, I will indicate how this document becomes a piece of evidence at the Supreme Court of Ghana when the minority NPP filed a petition contesting the results of the 2012 elections (Figs. 5.4 and 5.5).

The brief descriptions above indicate that the documents designed "participated with the situation" (Bitzer, 1992, p. 9) and thereby communicated the biometric technology to the electorates in positive (uncritical) terms. These technical documents provided a "fitting" response to the situation. In others words, the designers of the documents and online video represent what Kimball (2017) refers to as "institutional technical communicators;" a situation where communicators in an organization "become experts at representing how technology 'should' work, from the corporations perspective" (p. 3); and the technical documents become the medium through which the expert users represent knowledge about the biometric technology to the public. This is indicative of the fact that technical documents respond to an exigence. Below, I discuss how the documents maintained the logic of the biometric technology. I classify these documents as both system-constituting and system-maintaining (with specific reference to Seigel's characterization of technical documents) because they maintained assumptions and the logic of the biometric technology and also argued for why the biometric was necessary. More importantly, I look at how these documents "become members of an ecology of genres" (Spinuzzi, 2003, p. 62); that is, how they become an "interrelated group of genres used to jointly mediate the activities that allow people to accomplish complex objectives" (Spinuzzi & Zachry, 2000, p. 172). In essence, the documents worked together to articulate users into accepting the biometric device, while they open a window to articulate the biometric as a rhetorical artifact which has cultural significations.

While technologies move from one place to the other, local users design their own stories about the technology they adopt in technical genres to maintain or carry across the internal logic or ideologies that accompany those technologies. As a contribution to conversations about localization, this chapter examines how technical genres designed in Ghana by Ghanaians and for Ghanaians, used language to describe the biometric and how such use of language exposes the deterministic use of language to represent technology. It also reveals how local knowledge making about a technology in a decolonial contexts is masked in colonial/global rhetorics.

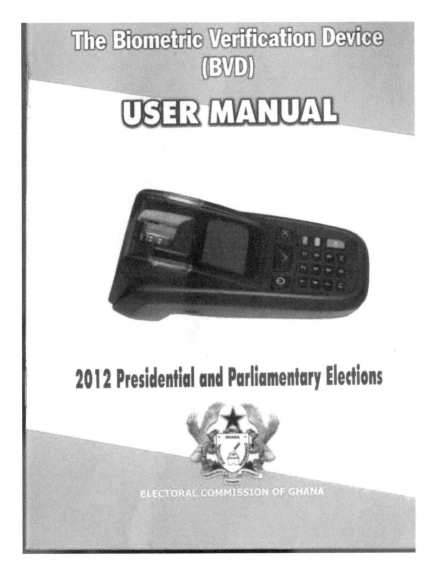

Fig. 5.4 Oblique overview of BVD

Fig. 5.5 Instructions showing outcomes of verification

Defining Linguistic Localization

In Chapter 1, I referred to Longo and Fountain's claim that technical communication is a system that orders, shapes, and produces knowledge about science and technology through the use of text. In this chapter, I add that technical documents do not only order, shape, or produce knowledge, but they also maintain and adapt knowledge about the technologies they accompany. I argue that one of the ways to maintain such logic and assumptions is the use of language to describe the instrumental features of the artifacts. And in most cases, technical documents are the means through which the logic of a technology is revealed or constructed. This is what I refer to as linguistic localization, that is, when language is used by local users to maintain the logic of a technology which has been adopted. The goal of such use of language is to help users adapt to the logic of the technological system. It is worthy to note that this use of language to describe the exceptional features of the biometric did not emanate from the designers of the technology but from the users of the technology, through locally designed genres. Thus, by virtue of the fact that the documents were designed locally, I argue that they capture "local logics, rhetorics, histories, philosophies, and politics" (Agboka, 2014, p. 298). Linguistic localization therefore works at the nexus between language, local–global culture and context of use. In subsequent paragraphs, I demonstrate how this kind of localization is exemplified in four different genres surrounding the biometric device: the user manual that accompanied the biometric device, parliamentary hansard, suite of posters, and an online video.

Linguistic localization as used in this chapter constructs the biometric as a powerful tool. To this end, words and phrases are chosen carefully to create the right impression of the biometric. Such use of language, to borrow from Bazerman, becomes the rhetoric that accompanies the biometric technology and makes it possible—it is the rhetoric that makes the biometric technology fit into Ghana's electoral system and makes the electoral system fit with the biometric technology. As I will indicate below, the language used to describe the biometric in the technical documents analyzed enables an examination of discourses and "patterns of thinking...concerned with the development, production, and marketing of artifacts and practices" (Miller, 1994, p. 92). These discourses also provide "a tool that facilitates meaning making about technological aspects of the world" (Koerber, 2000, p. 59). Linguistic localization, thus, provides an "affordance," defined by Norman (1988) to mean "the perceived and actual properties of the thing, primarily those fundamental properties that determine just how the thing could possibly be used" (p. 9). We come to understand that technologies are always already instrumental because we develop them to meet the needs of humans and they are always "part of human needs, desires, values, and evaluation, articulated in language and at the very heart of rhetoric" (Bazerman, 1998, p. 383). More so, technologies must go through several public for approval before they become a material reality. We can infer from discussions that linguistic localization is about discourses that encourage users to develop faith in a material object. I would want to add that not only is it about discourses that make technologies come to be, it is also thinking about the ways that we make these artifacts come to exist in the world. Bazerman, Miller, and Koerber make us understand that technologies must be accepted by intended users before they become a material reality. Hence, there must be arguments made for the value of these technologies at different levels. When such moments arise, accommodators fall on linguistic features to send the message across.

Linguistic localization emphasizes the functional features of a specific technology. For instance, the user manual provided procedures and instructions on how to use the biometric technology; the online video made a case for the efficient and credible nature of the biometric; the suite of posters communicated the potency of the biometric to present transparent results and capture impersonators; and the parliamentary hansard provided a legal mandate that gave the biometric teeth to bite perpetrators. Put together, these genres argue for why the biometric should be adopted and why it is a credible means of solving the country's electoral woes while they also maintained the logic of the biometric technology. These dual roles of the genres under discussion indicate that technical documents can: (1) maintain the logic of technology, and (2) constitute a technological system themselves. These blending roles indicate that it is difficult to give distinct roles to technical documents, that is, to say that one cannot construct a binary relationship between the roles genres play. It is not enough to say one document maintains the logic and the other constitutes a technological system. These roles are blurry. This adds to the fact that technical documents are not static; instead, they are fluid and dynamic and can play multiple roles. Although three different institutions, namely the EC, Parliament of Ghana, and STL, produced these documents, discourses were cast in three rhetorical strategies: identification of electoral malpractices, the urgent need to resolve such malpractices, and biometric as the savior of electoral woes.

'Scapes' and Flows: Transnational Flow of Biometric Technology

In *Modernity At Large: Cultural Dimensions of Globalization,* Arjun Appadurai uses the suffix "scapes" to speak to the fluidity of movements which have characterized the modern age; especially, the movement of technology, finances, media, ethnicity and ideology. Specifically, Appadurai (1996) uses "technoscape" to refer to the idea of technology movement around the globe. In this regard, technologies have become "fluid" or borderless and they move "at high speeds across various kinds of previously impervious boundaries" (p. 34). The Internet, for instance, can be considered a global technology as it has transcended traditional borders. The Internet technology has permeated the lives of people from poor countries and as well as rich countries. It has flattened the world, so to say. One other technology which is gaining global grounds at a supersonic level is the biometric technology. It is without a shred of doubt that biometric technology has become a very fluid technology which permeates every part of human society—its use is not limited to only one boundary or locale. As I stated in Chapter 2, biometric technologies are being used by states to secure national borders—hardly will anyone hold a passport which is not biometric verifiable, and hardly will people go through airport security without going through a biometric scan. Biometric introduction is part of the global effort to "shrink the 'identity gap'…which is seen as one of the primary obstacles to development" (Hobbis & Hobbis, 2017, p. 119) and in some instances, biometric is used to determine a foe or a friend, as in how it was used by the US to determine who was with Saddam or against him (Hristova, 2014, p. 516).

While technologies, such as the biometric, move across boundaries, they move with their specific internal ideologies, what loosely describes Appadurai's "ideoscapes." Appadurai uses this term to refer to the "concatenation of images…[that] have to do with the ideologies of states and the counterideologies of movements explicitly oriented to capturing state power or a piece of it" (p. 36). These ideologies which push forward the Enlightenment philosophy carry with them certain kinds of ideas, terms, and images. More importantly, ideoscapes are "constructed with a certain internal logic…" (p. 36). For example, a term like "democracy" comes with its internal logic which sustains it. Selfe (1999) for instance indicates that Americans have grand narratives that shape the culture of democracy: most Americans "have faith in the superiority of democratic systems and their conviction that the world would be a better place if democracy were spread around the globe. Within such a global system, all individuals, regardless of race, gender, creed, or sexual orientation, could exercise freedom of speech and religion; make choices about local, state, and national issues" (p. 116). Thus, democracy comes with such logic as freedom of speech and expression, equal opportunities and election of representatives in free and incontrovertible manner. Although Appadurai uses ideoscapes to refer to the movement of ideas such as democracy, freedom, and rights, I use the term loosely to refer to the ideologies that come with technologies that move from boundaries to boundaries.

It is significant that Ghana has also internalized the logic that comes with democracy and the processes that maintain the institution. Hence, the country adopts a technology which internal logic can help maintain the electoral system and instill confidence into electorates. This also means that the EC and stakeholders bought into and communicated the ideology of the biometric technology. The idea that technologies are ideological is not new. Several scholars before me have made reference to this claim. For instance, Banks (2006) identifies that "… all technologies come packaged with a set of politics: if these technologies are not inherently political, the conditions in which they are created and in which they circulate into a society are political and influence their use in that society and those politics can profoundly change the spaces in which messages are created, received and used" (p. 23). And Spinuzzi (2013) emphasizes the fact that technologies have ideologies which need to be maintained; hence, if one adopts a specific technology, the person adopts the ideologies that come with the technology (p. 117).

I indicated in Chapter 2 that the biometric comes with a specific sense of ideology: It is accurate, neutral, and objective and can authenticate and verify identities in a truthful manner. This is because of the belief that biometric collects "information that is only connected to you and no one else" (stlghana, 2012). In a way, these ideologies confirm Slack and Wise's (2005) assertion that "technologies in our culture are often linked with key values of the European Enlightenment of the eighteenth century: scientific objectivity, efficiency, and rationality" (p. 12). Even though Slack and Wise limit this technoculture to America, I believe these ideas about technology have moved to other countries. It is the reason we cannot discuss "technoscapes" without mentioning "ideoscapes" because as these technologies move, the ideologies move with them. The use of biometric by 34 low-to-middle-income countries, or by 20 countries in Africa (Makuilo) including Ghana, who have opted to utilize the biometric to enhance their electoral system because of the air of trust surrounding the technology is indicative of how nations around the world believe that they can live up to the Enlightenment notion of democracy by using technologies such as the biometric. Thus the biometric has become the means to an end.

The idea that technology is a means to an end has seen much discussion by philosophers and historians of technology. The basic underlying principle is that technology is a tool which can be used to achieve a certain goal. In this regard, technology, a mere tool, provides a quick fix to societal errors. For instance, the biometric is adopted to resolve electoral malpractices in Ghana. Embedded in this logic of quick fixes is the idea of *instrumentality*, that is, the extent to which things technical become part of human consciousness because of the perceived notion that they are neutral, objective, efficient, and they provide accurate results. At the heart of this idea of instrumentality is the belief that technology mediates human experiences in order to attain desired effects. Katz (1992) refers to this trust in things technical as the *ethic of expediency*, that is, a phenomenon where "science and technology become the basis of a powerful ethical argument for carrying out any program" (p. 264). When this happens, technology becomes both a means to an end and an end itself. The consequence of technological expediency is that technology becomes a form of consciousness, and our actions are exhibited in discourses that favor impersonality, closed-system thinking, narrow mindedness, and non-responsibility (Miller, 1978, p. 235). We also think in terms of efficiency and mathematization. More so, we are "enframed" and become "standing reserves," Heidegger expounds. As a matter of fact, Katz's ethic of expediency reflects Heidegger's (2014) notion of technology as a means to an end and human activity. In "The Question Concerning Technology," Heidegger intimates:

> The manufacture and utilization of equipment, tools, and machines, the manufactured and used things themselves, and the needs and ends that they serve, all belong to what technology is. The whole complex of these contrivances is technology. Technology itself is a contrivance—in Latin, an *instrumentum*. (Heidegger, 2014, p. 306)

Winner (1986) captures the instrumentality vividly by revealing that "...our lives are shaped by interconnected systems of modern technology, how strongly we feel their influence, respect their authority, and participate in their workings." Instrumentality becomes part of our psychological formation in which the technical aspect begins to control the rest of our individual personalities and we become conditioned to socially situate all problems in the realm of instrumentality and instrumental concerns. To think instrumentally, therefore, is to be trapped or "enframed" in the sense that every aspect of the human problem will be reduced to the logic of technology. It is not erroneous to say, therefore, that Ghana bought into the idea of instrumentation and subjected electoral malpractices to the logic of the biometric technology. To an extent, the biometric becomes the means that can affect a positive cause in the electoral system; it was going to help deliver efficient results to the people of Ghana.

This trust enshrined in the biometric and which was communicated in technical documents, such as the parliamentary hansard, suite of posters, user manual, and online video, reinforce the belief that documents accompanying a technology can perform one of three roles—they can either constitute a system, maintain a system, or disrupt a system (Seigel, 2013). The documents produced by Ghanaians maintained and constituted the biometric technology. System-constituting documents make a case for technology adoption: "they argue for a future establishment of a technology as a solution to an ideological, political or social problem" (Seigel, 2013, p. 41). These documents persuade users to adopt a technology to solve a problem. An example of a system-constituting document is the parliamentary hansard or the suite of posters designed by the EC. The suite of posters, for instance, states that biometric will capture and expose perpetrators while the hansard established that with the biometric, it was going to be difficult for people to use the information of others to vote. These statements, for me, are ways of reproducing the biometric ideology while they persuade Ghanaians to be confident in the new system.

System-maintaining documents, on the other hand, "keep users engaged with a particular technology" (Seigel, 2013, p. 71). To encourage users to constantly engage with a specific technology, such forms of documentation usually use imperatives to phrase commands and present task-oriented instructions. These styles are adopted to enable users "to quickly and efficiently complete the task at hand" (p. 71).

System-maintaining documents also provide users with the functional aspects of the technology in use. A typical example of a system-maintaining document is the user manual which accompanied the biometric technology. It is important to mention that while Seigel classifies system-constituting documents as deliberative because they concern future actions, system-maintaining documents are epideictic—they are concerned with present actions. In the coming pages, I demonstrate how the documents designed in Ghana either maintained or constituted the biometric ideology. I will also indicate the dangers involved in designing communication that solely maintains or persuades people to accept a specific technology and the implications for localization.

Local Documents Maintain the Logic of the Biometric

I maintain that in its quest to localize the biometric technology, the Electoral Commission designed both system-maintaining and system-constitutive documents. The implication is that the documents used to communicate the biometric technology—parliamentary hansard, suite of posters, user manual and online video—and by extension, technical writing, become the medium or instrument through which institutions, societies, organizations and individuals create exigence for a technology to be adopted. It also becomes a knowledge-producing practice. The user manual that accompanied the biometric technology, for instance, provides users with "knowledge and skills necessary" (Seigel, 2013, p. 41) to engage with the biometric in a way that voters can vote efficiently. This skill will prove to be insufficient as it does not instill in users the ability to think about the electoral system of Ghana and the role of the biometric in the general success of elections. Readers of the manual are presented with the "Functional Elements of the BVD (oblique view)" on page 2. The goal of this, I conjecture, is to give readers a general overview of the functional parts of the BVD so that they can know what to do and where to press when they want to perform a specific task at a specific time. It is not surprising that Seigel (2013) labels system-maintaining documents epideictic genres because they are more concerned with the present moment: how do I use the biometric device to help people vote? The user is therefore disengaged with the biometric immediately after elections are over. In this regard, system-maintaining documents maintain the logic of a technology through the use of instrumental discourses.

I use the logic of technology to refer to the ideology that technology provides quick fixes to problems that arise in social, political or communal contexts. The foreword to the user manual introduces the importance or purpose of the device: it was "introduced to the electioneering process in Ghana as a means of enhancing the integrity of the electoral process of the country" and it was going to "assist in the detection and prevention of practices of impersonation and multiple voting that have characterized our electoral process in the past;" the first page of the suite of posters reinforces the idea that biometric will "prevent incidents of multiple registration;" the online video establishes that "biometric registration is simply the collection of information that is only connected to you and no one else like your fingerprints and your photographs so that when we say it is you we know it is you." In addition, the rule that guides the use of the biometric system is revealed in the user manual: "a voter whose *fingerprints are not recognized by the device* will not be allowed to vote: while voters whose fingerprints are recognized by the device will be issued with ballot papers and allowed to cast the votes at the polling station." The highlighted phrase in the preceding sentence indicates a kind of agency given to the biometric. It is as though it's the voters' fingerprints that are flawed and the device is flawless. This is some kind of rationalized displaced agency that tries to encourage doubtless belief in the technology. Also, to introduce a technology to "enhance the integrity of the electoral process of the country" and "to assist in the detection and prevention of practices of impersonation and multiple voting..." exposes the trust the EC had in the technology. To an extent, an analysis of the manual exposes the belief that the use of biometric equals free and fair elections. This belief is unfounded and an unexamined assumption; it is rooted in the belief that technology is a means to an end, neutral and also an efficient way of solving an age-old problem in the electoral process. We are exposed to an "ideological complex based on a common sense belief in technology" (Selfe, 1999, p. xxii) and how such ideologies shape user or individual subjective identities. "Ideology," Althusser (1971) states, represents the imaginary relationship of individuals to their real conditions of existence. Mostly it refers to a system of representation and such systems of representations "always exist in an apparatus, and its practice, or practices" (Althusser, 1971, p. 28). These statements create, in the words of Steven Katz, a technological ethos. Thus, the machine becomes the mediator of truth and authenticity. What the foreword to the user manual and other documents accomplish is a barrage of promises about the successes that will be chalked when the technology is used. Little room is put into commenting on the quality of the biometric device. In this sense, the documents become an example of an epideictic documentation. By "epideictic" documentation, I mean to say that the manual is only "concerned with present action" (Seigel, 2013, p. 71). It provides "functional but not conceptual aspects" (qtd. in Seigel, 2013) of the biometric device. The sole purpose is to get users engaged with the biometric so that results are perceivably captured accurately and efficiently.

More so, although "The Biometric Device User Manual" does not explicitly say that if one fails to follow instructions to the letter the promise of delivering credible results will be compromised, this argument is implied in discourses surrounding the biometric technology. It is argued, for example, that the biometric device internally checks its own status and issues messages when necessary; or that the BVD gives an indication to the user when battery power is getting low. Relatedly, when BVD notices that the fingerprint scanner should be cleaned, it indicates. When such messages pop up, the user only has to either "replace the batteries;" "clean the fingerprint scanner" or take another action when necessary. These discourses implicitly push forward the promise that the device will enhance the integrity of the electoral process of the country by assisting to detect or prevent malpractices. The implication is that the system is foolproof; hence, any user of the technology can operate the BVD because the machine has in-built security mechanisms. But for a technological system to be established, must the logic only be maintained? The next section identifies that users must be convinced about the efficiency of a technology before its logic can be maintained. Hence, the need to develop a system-sustaining documents.

The representation of the biometric in the Parliamentary Hansard as a technology that has "teeth" (p. 791) which can bite or a technology which will make it "extremely difficult" for impersonators to vote is worthy of note. These linguistic representations exemplify Miller's notion of "technological forecasting," that is "a discourse in which the characterization and construction of moments in the present are crucial to the projection of the future." "The characterization and construction of moments," captured in rhetorical terms as *Kairos*, plays an important role in discussions about technology. As Miller indicates in "Opportunity, Opportunism, and Progress," *Kairos* provides an opportune moment for rhetors to "find an opening." Parliamentary deliberations provided a moment, an opening or a situation that needed a timely intervention. What was the moment that required intervention: there were "emerging challenges in the country's electoral system" (p. 785) and that it was necessary to adopt measures that will curtail the challenges. At this point, the electoral process becomes a risky endeavor. Defining the electoral process as at risk or the perception that elections in Ghana had experienced crisis moments provided a *kairotic* moment for the introduction of the biometric. The biometric, it was hoped, would provide a moment of change in the electoral process of Ghana. The biometric will provide intervention in the sense that it will make it "extremely difficult for a person to use the name and other particulars of another person to vote" (p. 788). Defining the moment, in this regard, also helps to identify the cause of the electoral woes of the country: People vote more than ones. Thus, with the biometric, the country will, ipso facto, seize the moment, and it will go through a credible and reliable electoral process that will yield accurate results. In this regard, the biometric becomes the "saving grace" of our electoral process.

To rhetorically establish the ethos of the biometric, some members discussed how

> with the introduction of the biometric registration and verification in the voting process it will be extremely difficult for a person to use the name and other particulars of another person to vote. (*Parliamentary Debates: Official Report [Emergency Meeting]*, p. 788)

It is gathered from the extract above that the country's electoral process suffers from impersonation, and the biometric becomes a way of solving this anomaly. The biometric offers a possibility because as the video which was used to educate the Ghanaian voter indicates, information about individuals are captured into a database for verification and identification. In the process of verification, a presiding officer is required to deliver "a ballot paper to a person to vote" (p. 786) after the officer has verified by "fingerprint or facial recognition that the person is registered voter whose name and voter identification number and particulars appear in the register." The new Constitutional Instrument stipulated that "voters shall go through a biometric verification process to cast their votes" (ibid.). By enacting this law, we encounter how the legislative body of the country gave power to the biometric to decide who may vote and who may not vote. This power enshrined to the biometric, in a way, speaks to the trust the legislative body had in the biometric. The biometric also becomes a powerful tool shaping the electoral activity in the country. Election 2012 was structured and organized differently from the previous elections held in the country. In later chapters, I will indicate how this law, contrarily, worked against the smooth running or the success of the elections—a revelation which indicates successful use of technology lies not in the enactment of laws to make technology fit into local activities but how these laws will be enacted to ensure that there is justice for all. The fear is that when we focus on enacting laws that will fit technology into local lives, we barely consider how the technology will be used by others to perpetrate acts that violate the laws that have been passed.

LOCAL DOCUMENTS SUSTAIN THE BIOMETRIC SYSTEM

System-sustaining documents, as I outlined above, work to "persuade a user to accept and establish" a technological system. According to Seigel (2013), such documents "argue for the… adoption of a technology, as a solution to…political or social problem" (p. 41). The genres under discussion helped to establish or persuade users to accept the biometric device. In the case of Ghana, the biometric technology becomes the means by which the country can record a free, fair, and incontrovertible elections. Kofi Ghana's statement in the video: "No biometric verification No vote;" or as it will be discussed later, the biometric will "expose" perpetrators. Ultimately, the biometric technology, "a techne, has become politeia" (Winner, 1986, p. 55), that is, the biometric has become political in the sense that it has become "a way of settling an issue in the affairs of a particular community" such as Ghana. Ghana's case provides two senses for us to discuss the political nature of the biometric technology. First, it intermingles with the human body in order that accurate information about individuals are gathered and stored for algorithmic codes. Second, it decides who may vote and who may not vote. This idea is captured in several ways. For instance, I revealed how operation officers were legally bound to turn away "a voter whose fingerprints are not recognized by the device…: while voters whose fingerprints are recognized by the device will be issued with ballot papers and allowed to cast the votes at the polling station" (manual, p. 1), and in the Parliament of Ghana, members of the House agreed that "voters shall go through a biometric verification process to cast their votes" (*Parliamentary Debates: Official Report [Emergency Meeting]*, 2012, p. 786).

It is also stated in the suite of posters that the biometric will "expose" perpetrators. The idea that the biometric will "expose" individuals provides a ground to interrogate the relationship between surveillance, biometric technology, and individual voters. What does it mean to be exposed in an election context? What are the consequences of the exposure and by what means will one be exposed? The goal of the "exposure," I believe, is to shame perpetrators and scare others away. One encounters notions of truth and authenticity in the rhetoric surrounding the biometric. The three stages that voters must go through enable the biometric technology to capture the image of individual voters. It is through the various images that have been captured that the biometric can prevent impersonators from voting because individual unique identities are established, coded, and stored. The stages of registration, outlined in the online video and scholarship on biometric technology, enable the biometric and voters to respond to two questions that form the heart of biometric technologies: Who are you? And are you who you say you are? The questions supposedly lead to a truthful identification of subjects who present themselves to the technology. These questions, in the words of Pugliese (2010), "underline the manner in which biometric technologies transmute a subject's corporeal or behavioral attributes into evidentiary data inscribed within regimes of truth" (p. 3). It becomes not just a part of an activity such as voting, it shapes the entire electoral and legal processes circumscribing the use of the biometric.

The biometric technology becomes a means by which bodies of individuals are managed and identified for the sake of accurate and transparent electoral results. In this regard, human behavior becomes something that can be "regulated, controlled, shaped" (Dean, 2010, p. 18) by the biometric technology and rules that have been established. These ideas expressed by statements made about the biometric technology exhibit tendencies of Foucault's concept of "governmentality". According to Dean (2010), governmentality entails two notions. One of such notions is the definition of "governmentality" as the "emergence of a distinctly new form of thinking about and exercising power in certain societies" (p. 28). This form of government is more concerned about populations, the basis for discussing biopolitics—the process in which state power is exercised over subjects or populations. For example during the 2012 elections in Ghana, the Electoral Commission, with its constitutional power to organize free fair elections, exercised power over the individuals by using a technology that could "expose" violators of electoral laws, "make it extremely difficult for a person to use the name and other particulars of another person to vote," and thereby "provide transparent" electoral results and "prevent incidence of multiple registration." In its quest to achieve this goal, the bodies of individuals were to go through verification and identification before they could cast their votes as citizens of Ghana. To this end, we can say that the biometric as used in Ghana did not only work to ensure a supposedly free and fair elections; among other things, it provided a regime of practice that enabled bodies to be organized, controlled, and even punished in some instances by the Electoral Commission and the government of Ghana.

Not only do we encounter a technological ethos, but we are also presented with particular forms of rationalities accompanying the biometric technology. Statements made about the efficiency of the machine expose "our understanding of technology as a way of solving social problems" (Selfe, 1999, p. 116) and confirms Slack and Wise's (2005) assertion that "technologies in our culture are often linked with... scientific objectivity, efficiency, and rationality" (p. 12). Though Slack et al. tell stories rooted in North American culture, this assumption about technology is revealed in a non-Western society. The biometric technology presents a kairotic intervention to the age-old problems bedeviling the electoral process in Ghana. These statements are forms of representing the technology as a worthy tool to be trusted: It is foolproof. Little consideration is put to discuss the political nature of technologies and how technologies such as the biometric system can fail. According to Magnet Shoshana, and as I will confirm in chapters six and eight, technologies such as the biometric technology are subject to failure. Shoshana Magnet (2011) maintains that "the complex communicative acts between biometric scanners, individuals, and institutions are, like any communicative exchange, filled with error" (p. 10).

In the online video, the biometric becomes an object of celebration. The biometric registration process becomes the means by which accurate information about voters and citizens can be gathered and stored for the purposes of verification, and more importantly a means by which electoral malpractices can be curtailed. In this regard, the designers of this online video, cast in the role of what I will term "technology accommodators," perform similar roles as technical communicators who transmit complex technological information to non-experts. Fahnestock (1986) identifies that authors of scientific reports are "overwhelmingly epideictic; their main purpose is to celebrate rather than validate" (p. 333) by making "all references to the amazing powers and secrets" of the biometric technology through "wonder" appeals or what Kitalong (2000) refers to as "magical" terms. Through the conversation among the three interlocutors, we can scrutinize the biometric "by picking apart the conjunction of the powerful discursive forces that create value for and give shape to technological developments and their uses" (Bazerman, 1998, p. 387). The biometric system becomes the embodiment of truth, accuracy, facticity, and a means of control. We could identify from the above representations that the biometric system really "affected the technical imagination" (Miller, 1994) of a cross section of Ghanaians so "powerfully."

Creating Biometric Ethos Through Chieftaincy

In the online video, the character, Kofi Ghana, exemplifies how a local authority is used to help persuade the Ghanaian voter about the efficiency of the biometric. Kofi is an Akan name for a Ghanaian male born on Friday, and his surname (last name), Ghana, gives an indication that he comes from Ghana. In most Ghanaian customary rites, it is believed that a child comes to the world with a birth name, which is literally the day you were born. The Akans of Ghana refer to this birth name as *Kra din*. "Kra" is the "soul" and "din" is translated as "name." Meaning that "kra din" is the "soul name" that every individual bear. You come into the world with the "kra din" before you are given any other name. I was born on a Wednesday, so my *kra din* is Kwaku. I use this as an example to indicate that "Kra din" is part of a Ghanaian identity. For me, the use of a character who bears a Ghanaian name and attribute is a way of localizing the biometric.

More so, he wears kente, a rich royal cloth worn only on special occasions by chiefs, a native sandal, and a crown. These are symbols of authority and kingship in a cross section of the Ghanaian society, especially those in the southern part of Ghana (I add that these days, *kente* is not a preserve for chiefs only; anyone who has the means can afford and use it). This adds that *kente* represents wealth and tradition. Through Kofi Ghana, we witness an attempt to accommodate technology through an elevation of one of the varied customs of Ghana, chieftaincy. A chief, as Utley (2009) posits, is an "amalgam of political head, religious head, judge, war leader, and advisor" (p. 64). Even though the chief's power has been lessened as a result of the practice of Western forms of democracy, he is still respected in the Ghanaian society. Thus, to use someone who epitomizes chieftaincy is in a way to first try to localize the technology and second communicate the credibility of the biometric technology. Here, Kofi Ghana, the epitome of the seat of traditional wisdom and power, adds what Aristotle calls "ethos" to the biometric. In *Rhetoric*, Aristotle outlined the modes of persuasion. The first mode of persuasion depends on the speaker's character; the second focuses on putting the audience into a certain frame of mind; the third provided by the words or logos of speaker (McKeon, 2009, p. 1329). In the first mode "persuasion is achieved by speaker's personal character when the speech is spoken as to make us think him credible" (McKeon, 2009, p. 1329). Aristotle believes that we believe good men readily; thus, speakers must be able to create impressions with what they say and not what people think of their character. Thus it was not mere happenstance that STL chooses a seat of authority as the accommodator of the technology to the electorate. When the speaker creates good character impression, it must stir the emotions of hearers (McKeon, 2009, p. 1330). As you can infer from the conversation above, the two listeners represented a cross section of Ghanaians who were confused about what it means to be registered biometrically. Ghanaians, at that point needed a more credible authority to listen to. At this instance, the best person to be listened to is the chief, an epitome of truth, power, and wisdom. The Akans of Ghana have a saying which goes "there are no bad chiefs, only bad advisors" (Utley, 2009, p. 67); this speaks to the trust and respect that Ghanaians hold for the traditional seat of authority.

Kofi Ghana also, in a way, is cast in the "expert/novice" binary that Johnson refers to in his book *User-Centered Technology*, and the two compatriots are the "idiots" who need to be dumbed-down with information about the biometric technology. The response they provide "wow, so it's really that simple!" could possibly speak to the fact that not only the American culture (referring to Johnson's case) celebrates technological idiocy but also the Ghanaian user of technology is captured in a similar metaphor. The voters in the video, possibly representing a cross section of Ghanaians, are represented as not having any knowledge of the modus operandi of the biometric; hence, they needed to "have everything provided to them in an easy-to-use format that asks little of them in return" (Johnson, 1998, p. 44). The interlocutors in the video, "not only present information and practical knowledge about [biometric] technology [as it is to be used in Ghana], but also creates, and sustains consumer's optimistic expectations" (Kitalong, 2000, p. 290). We become privy to the way the EC in Ghana and cross sections of Ghanaians established norms, values, and impressions about the biometric.

Finally, the use of phrases and words such as "very simple," "expose," "collection of information that is only connected to you and no one else" help us to understand how the ethics of technological expediency discussed by Steven Katz is employed in a non-Western context. Dwelling on Aristotle's concept of ethos, Katz makes us understand that when we perceive technology as an embodiment of objectivity, truth, and power, and thus make arguments for its adoption and use, then we fall prey to the ethics of expediency: "science and technology become the basis of a powerful ethical argument for carrying out any program" (Katz, 1992, p. 264). When this happens, technology becomes both a means to an end and an end itself. The impression we create is that technology is autonomous and deterministic, and it is "its *own raison d'etre* and driving force in culture" (ibid., p. 266). The consequence of technological expediency is that technology becomes a form of consciousness and our actions are exhibited in discourses that favor impersonality, closed-system thinking, narrow mindedness, and non-responsibility (Miller, 1978, p. 235).

Biometric as a Cyborg Technology

Here, I critically assess the relationship between the biometric technology and material lives of voters. I discuss the biometric as an example of a cyborg technology. I defined the biometric technology as a system which "scans a subject's physiological, chemical or behavioral characteristics in order to reify or authenticate their identity" (Pugliese, 2010, p. 2). This definition is well exemplified in Kofi Ghana's statement that the biometric is "simply the collection of information that is only collected to you and no one else like your fingerprints." These two definitions make an explicit statement about the biometric: It is a fusion between a part of the human body and a technical device. In other words, it is an intimate, complex, and dynamic relationship between the human body and a technology. The various stages of verification and authentication blur the boundary between technology and humans. According to Haraway (1991), a cyborg is a "cybernetic organism, a hybrid of machines and organism, a creature of social reality" (p. 149). A cyborg intermingles organisms and machines, "each conceived as coded devices" (ibid., p. 150). Hence, embodiment plays an important part in biometric discussions. In biometric systems such as the one Ghana used for the election, "the body…itself becomes a most engaging being" (ibid.; Haraway, 1991, p. 199). The biometric becomes what I term a "biocyborg" because it needs a part of the body to function and biological information, in turn, finds a permanent dwelling in the algorithms generated by the device: a symbiotic relationship, I would argue. It is through this symbiotic relationship that we encounter the political nature of technologies (topic for next section) or the various ways that technologies such as the biometric control.

The point, thus, is, cyborg ideas cannot be applied to only Western societies. In a globalized world where technologies are constantly being transferred from one context (usually Western societies) to the other (mostly unenfranchised cultural sites), such ideas as cyborg move along with the technologies. Thus, Ghana's case becomes a tacit point to look at how technologies that are adopted exhibit tendencies that are similar to or different from how biometric functions in Western society. Through the concept of cyborg technology, we encounter biometric as "the complex intersection of bodies, subjects, technologies and power" in both non-Western and Western societies.

Not only is the biometric a cyborg technology, but the voters are also implicitly interpellated to be part of the cyborg world: They are always already cyborg citizens. As part of its modus operandi, the biometric, as a cyborg, theorizes and maps the body "as a coded text whose secrets yield only to the proper reading conventions" (Haraway, 1991, p. 206). When the voter goes through the various stages of registration; when the voter's fingerprints are captured and when the biometric machine processes the captured image into codes, the voter's body is "thoroughly denaturalized" (p. 208); the body becomes "a coded text" that can be ordered, structured, and interpreted. More so, the body becomes a raw material that is "…appropriated, preserved, enslaved, exalted" (ibid., p. 198) and turned into codes by the biometric cyborg. The body is reified in the process. As we shall encounter in the next section, it is through vulnerable moments; through the process of reading and being read; scanning and being scanned; it is in these dualisms that technologies exhibit their control and their political nature.

Biometrics and the Politics of Technology

In the case of Ghana, the biometric technology becomes the means by which the country can record a free, fair, and incontrovertible elections. As Kofi Ghana tells us "No biometric verification No vote;" or as it will be discussed later, the biometric will "expose" perpetrators. Ultimately, the biometric technology, "a techne, has become politeia" (Winner, 1986, p. 55), that is, the biometric has become political in the sense that it has become "a way of settling an issue in the affairs of a particular community" (ibid., p. 22) such as Ghana. Ghana's case provides two senses for us to discuss the political nature of the biometric technology. First, it intermingles with the human body in order that accurate information about individuals are gathered and stored for algorithmic codes. As we have realized from the preceding discussion, "if there is no body, then there can be no biometric as such," (Pugliese, 19). In this regard, through an analysis of the biometric system, we can even diffuse the notion that there is a binary between body and technology. There is, as Pugliese states, "an inextricable relation between bodies and technologies" (p. 20). Second, it decides who may vote and who may not vote. This idea is captured in several ways. In Chapter 4, for instance, I revealed how operation officers were legally bound to turn away "a voter whose fingerprints are not recognized by the device…: while voters whose fingerprints are recognized by the device will be issued with ballot papers and allowed to cast the votes at the polling station" (manual, p. 1). In Kofi Ghana's narrative, we are told that "no biometric verification, no vote;" and in the Parliament of Ghana, members of the House agreed that "voters shall go through a biometric verification process to cast their votes" (*Parliamentary Debates: Official Report [Emergency Meeting]*, 2012, p. 786). In essence, the biometric assumes a very political position. It becomes not just a part of an activity such as voting, it shapes the entire electoral and legal processes circumscribing the use of the biometric. To this end, we can discuss the biometric not only as a manifestation of a cyborg technology, but also, in terms of *biopower*.

By *biopower*, I draw on Foucault's (2009) concept of biopolitics referenced in his lectures on security, territory, and population as a "set of mechanisms through which the basic biological features of the human species became the object of a political strategy, of a general strategy of power, or, in other words, how starting from the eighteenth century, modern Western societies took on board the fundamental biological fact that human beings are species" (p. 1). And exemplified by Haraway as "the practices of administration, therapeutics, and surveillance of bodies that discursively constitute, increase, and manage the forces of living organisms" (p. 11). Biopower, for Foucault becomes a technique used to control people or populations. It is in a sense the ability to exercise power over bodies so that truth about individual voters is captured. This way, it is believed, results will be transparent because there will not be multiple voters.

Even though Foucault's *biopower* makes explicit reference to Western societies, discourses generated about the biometric in Ghana, a non-Western society, make tacit reference to ideologies inscribed in Foucault's biopolitics. For instance, Kofi Ghana tells us that when voters go through the biometric verification process, "information collected" will be connected to "you and no one else." Hence, the quest to achieve a free fair elections will be met and results will be "transparent." In this direction, the biometric becomes a technology that enhances truth and accurate results. In Chapter 4 and later in this chapter, I refer to this belief in the biometric as an ideology. Thus, I refer to the biometric as a biopolitical technology because it served as a mechanism that restricted miscreants from voting or registering twice. The biometric ensures "one man one vote." I must also say that even though the biometric could be used differently at different places, its use no matter the context, establishes a truism: The binary between technology and humans is becoming blurred, and increasingly humans are becoming over-dependent on technology.

Biometric as the Articulation of Gendered Patterns[2]

Ghana's adoption of the biometric helps to advance a case that increasingly, as technologies such as the biometric are advanced, societies will depend solely on these technologies to explain or define who they are and what counts as truth. Hence, technologies such as biometrics will be able to speak humans and assign meaning to who we are or what we are (Koerber, 2000, p. 66). An example of technology-speaking human can be identified in the processes voters in Ghana would have to go through before they cast their votes. As I stated in Chapter 4, one of the major laws that backed the biometric stipulates that voters can only vote when biometric recognizes the fingerprint of the voter. This is exemplified by the stages laid out by Kofi Ghana in the video I analyzed and made more concrete by Parliament's institutionalization of a Constitutional Instrument that ensured that "voters shall go through a biometric verification process to cast their votes" (*Parliamentary Debates: Official Report [Emergency Meeting]*, 2012, p. 786). If the biometric fails to recognize finger, that individual is not allowed to vote. Thus, the biometric becomes a speaker of an individual voter's identity. It must confirm that you are you before you can vote. These practices of going through different stages to be scanned, identified, verified, and registered are forms of what Scott calls "disciplinary rhetorics." These disciplinary rhetorics work to shape "people as particular kinds of subjects and subjects them to various exercises of power" (Scott, 2003, p. 7). This is a notion I elaborated on in my previous section and will return to as I advance my argument in this chapter.

If we are becoming dependent on technology, and if technology speaks about us, through verification and identification, then Koerber (2000) admonishes us to pay attention to the various ways that technologies become "speakers of meanings assigned to…gender" (p. 66); a reiteration of Durack's (1997) assertion that technology may have gendered meanings. I state that we could, somewhat, read a gendered meaning into the biometric usage in Ghana. As is stated in the video, the biometric ensures that "one man one vote that counts." Why should it be "one man, one vote" and not a different statement? Are women not allowed to vote in Ghana? Perhaps, we may say that this statement was made because of a cultural blinder that makes it difficult to filter the various ways that discourses about technology or technology use come pregnant with gendered meanings. It could perhaps be an example of what Durack terms an "unintentional bias" or a neutral statement, but no matter the direction we take it, whether it is literal or not, the phrase confirms statements by some feminist scholars that technologies are gendered and mostly come clothed in masculine terms (Durack, 1997; Haraway, 1991; Koerber, 2000; Wajcman, 1991). This statement reflects Koerber's (2000) elaboration of critiques feminists have leveled against technology: First, technology is intrinsically linked to masculine ideology; second, technology shapes the way we think about gender and humanity; and third, technology is perceivably neutral and objective (pp. 63–68); and Wajcman's (1991) claim that "the very language of technology, its symbolism is masculine" (p. 19). This intricate relationship,

Wajcman argues, is not inherently biological but it is as a result of how gender has been culturally and historically constructed. As innocent as the statement made by Kofi Ghana may be, it is, I would argue, one of the instances where the role that women play in technological or political discourses are sidelined. The man is technological, the woman, a passive, private, sidelined consumer of an artifact.

Even though these feminist arguments are made in regard to the relationship between gender and technology in Western societies, I make a case that it becomes relevant in a non-Western society because it provides avenues for scholars to interrogate the various ways that technologies that are adopted by non-Western societies carry the stigmata of Western cultural ideologies. It paves a way to, as a matter of urgency, investigate the ways "gender affects both access to technology and the practices of users of technology" (Wells, 2010, p. 151) in a non-Western society. We can also begin to interrogate how women in non-Western societies, such as Ghana, have been represented in technological and democratic discourses. The ultimate question to ponder over is does the representation of the biometric as a technology which ensures "one man one vote" reflect the cultural, historical, rhetorical, and political relationship that has existed between men, women, and technology in the Ghanaian society in general? Does this phrase reinforce Wajcman's (1991) claim that technology "fundamentally embodies a culture?" (p. 149). Or does it reflect statements made by Wajcman (1991) and Reed (2014), for instance, that technologies do carry imprints or biases of their creator? Since the biometric was not developed in the country, can we make a case that the device carries a bit of the cultural bias of the production country? To an extent, I will respond to these questions I pose in the affirmative. The idea that the biometric ensures "one man one vote" cannot be taken lightly. It provides a way for technology accommodators to think critically about the language they use to represent technology to users or customers.

The statement I have been commenting on, "one man one vote that counts," provides an avenue to discuss gender, technology, and democratic culture in Ghana because even though the call for the participation of women in national governance began half a century ago, response to this call has been snail paced. For instance, the number of women in Parliament decreased from 25 in 2008 to 19 in 2012. This retrogression made Oquaye, for example, wonder what the number will be in 2016 (Oquaye, 2012). This call for women to be part of democratic processes is on the ascendancy in Ghana because even though the participation of women in "national governance" is "a requirement of democratic governance and a condition for achieving the Millennium Development Goal" (Manuh, 2011, p. 1), their participation still linger below the minimum 30% the UN system has proposed. Factors such as "increasing monetization of politics and political campaigns, and growing political violence and intimidation, creates insecurity for women and builds resistance to their participation..." (Manuh, 2011, p. 2). According to Oquaye (2012) "military intervention in politics inflicted a lethal wound on women's political emancipation as the process of development through political parties was arrested in the face of machoistic male-dominated militarism;" a historical incident which worked to prevent women from fully participating in the public sphere. This has led scholars to propose several short-term or long-term goals to mitigate development. One of such affirmative actions is the quota system which requests that a number of seats should be reserved for women in parliament (Oquaye, 2012).

The point of my argument is that political, religious, and social organizations can garner all the resources and legal amendments they could but if discourses are couched in masculine terms, the woman could be sidelined. Kofi Ghana helps us to identify that it is not enough to adopt a technology and it is paramount to avoid representing technologies in masculine terms. If discourse about technologies that are to enhance democracy, such as in Ghana's case, is couched or discussed in masculine terms, we articulate and rearticulate incidents that prevent the participation of women in Parliament. It is not surprising therefore that an affirmative action policy enacted by the opposition NPP to encourage the participation of women and to allow female MPs in the party to contest for national elections unopposed was met with disaffection from, mostly, some male aspirants. The consequence of the disaffection was the redraw of the policy (MyJoyOnline, 2015). I contend that these oppositions to affirmative actions will be encountered the more when we do not work to counter and silence the representation of elections in Ghana as the preserve of males. I agree with Ocquaye (2012) when he proposes that "...the perception that politics is the realm of men should be removed by conscious efforts in advertising, symposia, debates, plays, films, etc" (p. 6). The representation of the biometric as a technology that ensures "one man one vote," I contend, should not be encouraged in the domain of politics or democracy in Ghana. The advert, though it informs the populace about the biometric, also implicitly carries an agenda that pushes the idea that technology use or elections in Ghana is masculine. This advert, I propose, must not be used in any of the Electoral Commission's information sessions; neither should it be encouraged by any organization that educates the electorate on electoral processes. It carries with it a rhetorical and cultural bias that reinforces gendered politics, a force that works against the participation of women. Hence, we can argue that the biometric technology carries the cultural bias of its creators and those who adopt them.

It is interesting to note that these local forms of representing the biometric is not different from global conversations about the biometric technology. Biometric is represented in instrumental terms by biometric technology companies. A typical example of instrumental representation is provided by Genkey. This biometric production company indicates that the biometric technologies they produce "are helping to deliver fair and transparent elections, securing the fundamental democratic principle of 'one person, one vote'" (Genkey). More so, their biometric products offer: "highest quality biometric registration of citizens;" and "fast and reliable de-duplication and identification," which means every identity given to an individual is unique. Smartmatic,[3] another biometric technology company, also acknowledges the biometric they produce instill "transparency," "integrity," and "efficiency" into election processes. More important, their biometric technology gives voice to voters and ensures that voters cast their votes "independently." While these forms of representation indicate how biometric is being used as a tool to maintain the logic of democracy, they also enable an understanding of how local knowledge making about technology in decolonial contexts is masked in colonial/global rhetorics. Is it that Ghanaian users knew about the efficiency of the biometric and communicated it to the public or they only reproduced the biometric ideology as was represented by biometric production companies? Whatever the case is, it is important to realize that instrumental discourse plays a major role in the communication of technology. Such discourses maintain and sustain interests in a technological system.

Through the documents analyzed, we identify that the biometric is articulated in a matrix of totalizing metaphors: success, hope, progress, truth, accuracy, efficiency, credibility and simplicity. Five threads of arguments extol the powerfulness of the biometric. The technology is articulated as:

- a very simple system to use;
- a credible source of information because it collects data that is connected to one person and no one else;
- the surest solution to the emerging challenges in the electoral system;
- a technology which can prevent incidents of multiple voting; and
- a technology which is able to expose perpetrators

Thus, it is most likely that a non-Western society willing to adopt a technology to solve specific problems, in my case, electoral problems, will draw on an aspect of its culture to develop rhetorical patterns that will persuade. Technical writing will be the powerful medium used to pursue this agenda. But what are the dangers involved in using totalizing phrases to capture technology? Do biometric technologies work as efficiently and accurately as communicated by the EC officials and biometric industries? Should we only concentrate on discussing instrumental features of technologies? We turn to the next chapter for a response.

Notes

1. http://www.parliament.gh/publications/.
2. I owe the development of this section to Dr. Gerald Salvage. When he reviewed a paper I submitted to a journal for publication, he pointed out how the phrase "one man one vote" can lead to the discussion of the biometric technology and gendered nature of democracy in Ghana.
3. https://www.smartmatic.com/.

References

Agboka, G. Y. (2013). Participatory localization: A social justice approach to navigating unenfranchised/disenfranchised cultural sites. *Technical Communication Quarterly, 22*(1), 28–49. https://doi.org/10.1080/10572252.2013.730966.

Agboka, G. Y. (2014). Decolonial methodologies: Social justice perspectives in intercultural technical communication research. *Journal of Technical Writing and Communication, 44*(3), 297–327.

Althusser, L. (1971). Ideology and ideological state apparatuses (notes towards an investigation). In L. Althusser (Ed.), *Lenin and philosophy and other essays*. London: New Left Books.

Appadurai, A. (1996). *Modernity at large: Cultural dimensions of globalization* (Vol. 1). Minneapolis: University of Minnesota Press.

Banks, A. J. (2006). *Race, rhetoric, and technology: Searching for higher ground.* Mahwah, NJ and Urbana, IL: Lawrence Erlbaum; National Council of Teachers of English.

Bazerman, C. (1998). The production of technology and the production of human meaning. *Journal of Business and Technical Communication, 12,* 381–387.

Bitzer, L. F. (1992). The rhetorical situation. *Philosophy & Rhetoric, 25,* 1–14.

Dean, M. (2010). *Governmentality: Power and rule in modern society.* London: Sage.

Durack, K. T. (1997). Gender, technology, and the history of technical communication. *Technical Communication Quarterly, 6*(3), 249–260.

Esselink, B. (2000). *A practical guide to localization* (Vol. 4). Amsterdam: John Benjamins Publishing Company.

Fahnestock, J. (1986). Accommodating science: The rhetorical life of scientific facts. *Written Communication, 3*(3), 275–296.

Foucault, M. (2009). *Security, territory, population: Lectures at the Collège de France 1977–1978* (M. Senellart, Ed., Vol. 4). New York: Macmillan.

Gonzales, L., & Zantjer, R. (2015). Translation as a user-localization practice. *Technical Communication, 62*(4), 271–284.

Haraway, D. (1991). *Simians, cyborgs, and women: The reinvention of women.* London and New York: Routledge.

Heidegger, M. (2014). The question concerning technology. In R. C. Scharff & V. Dusek (Eds.), *Philosophy of technology: The technological condition: An anthology* (2nd ed., pp. 305–317). West Sussex: Wiley Blackwell and Sons.

Hobbis, S. K., & Hobbis, G. (2017). *Voter integrity, trust and the promise of digital technologies: Biometric voter registration in Solomon Islands.* Paper presented at the Anthropological Forum.

Hristova, S. (2014). Recognizing friend and foe: Biometrics, veridiction, and the Iraq War. *Surveillance & Society, 12*(4), 516.

Johnson, R. (1998). *User centered technology: A rhetorical theory for computers and other mundane artifacts.* Albany: State University of New York.

Katz, S. B. (1992). The ethic of expediency: Classical rhetoric, technology, and the Holocaust. *College English, 54*(3), 255–275.

Kimball, M. A. (2017). Tactical technical communication. *Technical Communication Quarterly, 26*(1), 1–7. https://doi.org/10.1080/10572252.2017.1259428.

Kitalong, K. S. (2000). "You will": Technology, magic, and the cultural contexts of technical communication. *Journal of Business and Technical Communication, 14*(3), 289–314.

Koerber, A. (2000). Toward a feminist rhetoric of technology. *Journal of Business and Technical Communication, 14*(1), 58–73.

Magnet, S. (2011). *When biometrics fail: Gender, race, and the technology of identity.* Durham, NC: Duke University Press.

Manuh, T. (2011). Towards greater representation of women in national governance. *Governance Newsletter,* 17. Accra.

McKeon, R. (2009). *The basic works of Aristotle*. New York: Random House LLC.
Miller, C. R. (1978). Technology as a form of consciousness: A study of contemporary ethos. *Communication Studies, 29*(4), 228–236.
Miller, C. R. (1994). Opportunity, opportunism, and progress: Kairos in the rhetoric of technology. *Argumentation, 8*(1), 81–96.
Miller, C. R. (1998). Learning from history: World War II and the culture of high technology. *Journal of Business and Technical Communication, 12*(3), 288–315.
MyJoyOnline (Producer). (2015, November 21). NPP affirmative action scrapped; November Congress to Consider Adoption.
Norman, D. A. (1988). *The psychology of everyday things*. New York: Basic Books.
Oquaye, M. (2012). *Reserving special seats for women in parliament: Issues and obstacles*. Accra: Governance Newsletter.
Parliamentary Debates: Official Report (Emergency Meeting). (2012). Retrieved from Parliament House, Accra.
Pugliese, J. (2010). *Biometrics: Bodies, technologies, biopolitics* (Vol. 12). New York: Routledge.
Reed, T. V. (2014). *Digitized lives: Culture, power, and social change in the Internet era*. New York: Routledge.
Scott, B. (2003). *Risky rhetoric: AIDS and the cultural practices of HIV testing*. Carbondale: Southern Illinois University Press.
Seigel, M. (2013). *The rhetoric of pregnancy*. Chicago: University of Chicago Press.
Selfe, C. L. (1999). *Technology and literacy in the twenty-first century: The importance of paying attention*. Carbondale, IL: SIU Press.
Slack, J. D., & Wise, J. M. (2005). *Culture and technology: A primer*. New York: Peter Lang.
Spinuzzi, C. (2003). *Tracing genres through organizations: A sociocultural approach to information design* (Vol. 1). Cambridge, MA: MIT Press.
Spinuzzi, C. (2013). *Topsight: A guide to studying, diagnosing, and fixing information flow in organizations*. CreateSpace Independent Publishing Platform.
Spinuzzi, C., & Zachry, M. (2000). Genre ecologies: An open-system approach to understanding and constructing documentation. *ACM Journal of Computer Documentation (JCD), 24*(3), 169–181.
stlghana. (2012). *STL Ghana—Biometric system voter registration process*.
Utley, I. (2009). *The essential guide to customs and culture: Ghana*. London: Kuperard.
Wajcman, J. (1991). *Feminism confronts technology*. Pennsylvania: The Pennsylvania State University Press.
Wells, S. (2010). Technology, genre, and gender. In S. A. Selber (Ed.), *Rhetorics and technologies: New directions in writing and communication*. Carolina, SC: The University of South Carolina Press.
Winner, L. (1986). *The whale and the reactor: A search for limits in an age of high technology (La Baleine et le réacteur)*. Chicago: Chicago University Press.

CHAPTER 6

User-Heuristic Experience Localization

In the preceding chapter, I examined the super-hero narratives used to represent the ethos of the biometric technology in technical genres: it is efficient, objective, fast and value-neutral, it will expose perpetrators, work perfectly and capture or collect data on individual voters accurately so that "when we say it is you, we know it is you." These forms of representation, Magnet Shoshana (2011) indicates, are undergirded by the scientific notion that "the human body can be made to speak the truth of its identity through the use of biometric technologies" (Magnet, 2011, p. 3), or that the biometric plumbs "individual depths in order to extract their core identity" (p. 5). The technology is represented as "a fail-safe solution to all our problems" (p. 8) and biometric provides "mechanical objectivity" (p. 5).

This chapter, thus, presents a counter-narrative to the positive rhetorical representation of the biometric technology and argues for why user interaction with technology plays a major role in localizing a technology. Biometric technologies are not "fail-safe" as rhetorics surrounding the technology suggest. Evidence of biometric failures abound. As I have indicated throughout this book, the biometric technology used by Ghana failed in some polling stations at a point during the electoral process. Let me refresh our memories with a quote from Mr. Gadugah:

> Then come election day it broke down, some people couldn't use it, some people had to use the manual registration which was outside the law and in fact some people got disenfranchised because the machines broke down…

> When you take the verification device, for example, the printers were just breaking down like that because they could not take the pressure. If you start printing, you print 1, 2, 3, 4 then the printer breaks down… the BVD failed and STL, the technicians, also blamed it on humidity, high temperature.

Technical reports which recorded election proceedings also confirmed claims about breakdowns. The Coalition of Domestic Election Observers (CODEO) reported that the biometric machine failed "at some point at 19% of polling stations during voting" (*Final Report on Ghana's 2012 Presidential and Parliamentary Elections*, 2013, p. 3); and during the 2016 voter registration, they observed that the biometric failed to function in 6% of the centers they visited (Arhin, 2016, p. 1). Thus, instead of seeing the biometric as the savior of Ghana's electoral system, these revelations introduce an important conversation about localization. It broadens our perspective to think about the role that weather conditions play in technology adoption processes. Moreover, Hobbis and Hobbis (2017) identified that when the biometric technology was deployed in Solomon Islands during the 2014 election, the technology rejected individuals engaged in "slash-and-burn agriculture, or non-industrial fishing" and those whose faces have gone through some form of medication "from aging through the application of cosmetics, plastic surgery or …tattoos" (Hobbis & Hobbis, 2017, p. 118). They argued, based on the observation in Solomon Islands, that "whatever the weaknesses and strength of the digital BVR system may be, it is only the ***user experience*** that tells the full story of its implementation, use and eventual effect on the electoral process" (p. 119). I agree.

From User Experience to User-Heuristic Experience

These stories about biometric failures help to confirm the truism in Hobbis and Hobbis' call for more attention to user experience, defined to mean "all aspects of the end user's interaction with the company, its services and its product" (Norman & Nielsen, 2008). In this definition provided by Nielson and his group, we encounter an intricate relationship between the company, its services and its product. So, in Ghana's case, for instance, we are thinking in terms of a relationship between the company which produced the biometric technology (unknown), the services they provide and the product (biometric technology). But I indicated in Chapter 1 that none of the EC officials I interviewed could tell me the name of the company which designed the biometric used. If they could not tell me the company, then I could argue that the EC did not enjoy the services of the biometric designers either. We are only left with one item on the list of relationships, that is, the product (the biometric). This break in the communication channel between a company and its users is what localization scholars have been working to address. This gap also implies that sociocultural contexts surrounding the use of the biometric technology is ignored by the designers of the biometric technology. What happens when users did not enjoy any relationship with the company or its services? Alternatively, how can designers take into consideration the local contexts of use when they did not have any contact with the end user? How do forces of globalization shift or shape the definition of user experience provided by Nielson and his group?

Another good definition of the term "user experience" is provided by User Experience Network: "the quality of experience a person has when interacting with a specific design." In this definition, there is much more focus on the user than there is on the relationship between company, service and product. Yet still, this definition is limiting as it emphasizes only the experience and not the action users initiate to make the technology successful. So, going by these two definitions we can create this scenario: I encounter a technology like a mobile phone, a car, or a cup. This meaning of merely encountering a technology is implicit in User Experience Network discussion of user experience when they give the range of encounters to be "specific artifact, such as a cup, toy or website, up to larger, integrated experiences such as a museum or an airport" ("User Experience Network," 2008). What was my experience when I encountered these technologies? The respond can be good experience, bad experience or it wasn't that great. When we focus more on just the mere experience, we might be tempted to collect data only on positive or negative encounters with a technology without paying attention to user actions. In essence, we might miss the opportunity to ask how one managed to use the technology when the encounter with the technology was not the best. This means that we might not focus on the creativity of users or the people who encountered the technology. Sun (2012) reveals that "missing the actual practice of social activities is a big problem in cross-cultural design" (p. 31) and argues for a technology design "that takes user experience as both situated action and constructed meaning" (p. 31). This has two implications for us. (1) when technology design is decontextualized, we can only assess the end user's interaction with the product, and (2) we need to expand on the definition of user experience to emphasize on actions that users take to the neglect of the other factors: company and services. I have indicated that a call on attention to users and their actions in localization processes has received much attention in the field of technical communication. Sun (2012) calls the efforts by user's *user localization*. She identifies elsewhere that user localization is both a product and a process. As a product, localization indicates the intricate relationship "with a locale" (Sun, 2009, p. 258) and the cultural and social factors that emanate from that culture; and as a process, localization indicates user interactions that enable the integration of a technology into a user's culture. User-heuristic localization, as I flesh out in this chapter, extends on Sun's notion of localization as a process.

Therefore, for the sake of the argument I seek to make, and for the sake of the localization process I want to advance, I focus on users experience of the biometric technology and identify that a tension exists between the representation of the biometric technology, implementation and use in some quarters where it was deployed. I believe that if we pay much attention to user experience, we will come to know that users have tacit forms of knowledge that they bring to technology. This gap which usually ends in mass breakdowns of the technology provides an avenue for users to innovate or strategize to ensure that the technology they adopt is used effectively. In the case of Ghana, we focus on how users used tactics and strategies to ensure that biometric use did not jeopardize the electoral process. Users' interaction with the biometric will unravel decisions and actions that "undermine the theoretical potentials or promises of 'anti-corruption' technologies" (Hobbis & Hobbis, 2017, p. 115) such as the BVD used by Ghana. Moreover, users' interactions indicate that users are innovative users who initiate strategies to integrate technology into their local complex contexts. For localization scholars, a focus on user strategies "allows us to move beyond technologically deterministic understandings that assume the (effective) deployment of a given technology will lead to a particular outcome (improved voter integrity and possibly a more stable electoral and political system)" (Hobbis & Hobbis, 2017, p. 115).

As I indicated in Chapters 1 and 5, the biometric technology adopted broke down on several levels: (1) biometric performed poorly because it could not withstand the heat in Ghana; (2) the machine could not read the fingerprints of some voters; (3) training in biometric use didn't really help since most users struggled to use the technology on Election Day; and

(4) User instruction manual was confusing. There is every indication that users were frustrated with the integration process, and even with the EC; but on a more theoretical level, the breakdowns reinvigorate broader concerns by localization experts that technologies come packaged with assumptions that do not reflect or support user needs in every new environment they encounter. Hence, users are frustrated with new technologies. The image of a "frustrated user" of technology pervades conversations about technology use. Johnson (1998) for instance, states instances where a wristwatch manufacturing company realized that users were returning a significant number of the watches purchased or a fax machine production company realized through survey that 95% of users who bought the fax machine never used them (p. 69); and Norman (2013) also narrates a story of his friend who "got trapped in the doorway of a post office in a European city" (p. 2). This friend tried several heuristic approaches to get out of the "iron cage" but to no avail (for full details of the image of the trapped man and how he struggled with the door, see the Design of Everyday Things). Even though it is not explicitly stated, one would conjecture that the wristwatches were being returned or the fax machines were not put to use because the users were in a way frustrated with the technology in their possession. Labeling this scenario as the "breakdown between humans and technological artifact" (p. 70), Johnson reveals that two issues lead to such breakdowns: the first is the idea that system experts are the well-trained brains to make decisions concerning the design of a technology, and second, systems are designed to "reflect rational decisions" (p. 70); Don Norman blames this on the existence of two separate gulfs: the execution gulf and the evaluation gulf (38), and Huatong Sun (2009) blames it on a gap that exists between designers culture and users culture.

The good thing is that Johnson and Sun believe that the user can step into bridge gaps while Norman indicates that "the role of the designer is to help people bridge the two gulfs" (p. 38). Ironically, Norman uses an example which portrays the user as a hero who can save technology from causing enough frustrations. He cites an example of an elderly woman who struggled to open a filing cabinet but could not do so. He offered to help this lady and he also tried various heuristic approaches before this cabinet could open. Norman, wiggled the drawer, twisted, pushed hard and banged it to open. But he asks an interesting question which leads to user agency: "what other operations could be done to open the drawer?" (38). At this point of questioning, the user had exhausted all meaningful approaches to getting the technology to work but nothing happens.

Defining User-Heuristic Experience

For me, at the level of exhaustion is when users resort to heuristic approaches to solve the technological crisis. This is an indication that users "can also design workarounds, ways of overcoming the flaws of existing flaws" (Norman, 2013, p. xii). The point is that frustrated users are not helpless, agentless users; at the point of frustration, they figure out how to save a technology. And rightly so, Ghanaian users devised strategies to salvage biometric breakdowns. Although officials at the Electoral Commission could not make changes to the design of the biometric, they made decisions that sought to influence the sociocultural factors affecting the use of the biometric by resorting to the use of canopies, a decision Mr Gadugah believes further escalated the breakdowns. Not only did the EC officials use canopies to control temperature; they also devised

> many unscientific methods in order to enable the machine capture the fingers of some prospective voters. *Some people had to buy Coca Cola (soft drink) to wash their hands* before they were captured at the Inti Suariya Primary School polling station. In the Kanvilli R/C Polling Centre in Tamale, *a hot coal pot was provided for voters to dry their hands after washing with soap and water before their fingers* could be captured. Most of those affected by the problems of verification *were elderly women, some young women who had dyed their hands and fingers with a local herb called "lenle"* were also affected. (CHRAJ report, p. 6)

It is interesting that the report written by the Commission for Human Rights and Administrative Justice (CHRAJ) classifies measures devised by users to solve the situation as "unscientific." This statement recasts conversations about how knowledge is classified in our societies. Since the time of the ancient Greeks, this categorization of knowledge has permeated our societies. By referring to user methods as "unscientific," the report writers cast heuristic approaches adopted by users within the domain of the temporary, contingent and non-expert form of knowing; a kind of knowledge associated with users, and it also supports the notion that users live in the world of the ordinary. More importantly, by calling approaches "unscientific," report writers indicate that such forms of knowledge classification are not limited to Western cultures as purported by scholars such as Johnson; but instead, it is a global issue. Now that technology is transferred to different parts of the world, it will be appropriate to study the various ways user knowledge is devalued and unrecognized.

It does not come as a surprise that users ways of solving biometric breakdowns were labeled "unscientific" because as Johnson (1998) has revealed, users of technology are perceived to be living in "the land of the mundane," that is, a domain which is associated with unscientific, "common, ordinary, or, of this world" (cited in Johnson, 1998, p. 3); and the knowledge of the commoner has become voiceless, knowledgeless, and it is colonized by expert knowledge. Barton and Hamilton (1998) refer to this kind of mundane knowledge, vernacular knowledge or vernacular practices. According to these scholars, vernacular practices "are essentially ones which are not regulated by formal rules and procedures of dominant social institutions which have their origins in everyday life" (p. 249). Like Johnson, Barton and Hamilton identify that less value is placed on vernacular practices because such forms of knowledge are not acquired in the classroom or through formal education. According to these literacy theorists, vernacular knowledge has two main characteristics: first, it is learned informally; meaning that, such kind of knowledge is "not systematized by an outside authority" (p. 252). The role of the learner is not fixed but it varies from context to context; second, in vernacular knowledge, learning and use are integrated. Unlike in the educational setup where people learn through an apprenticeship or when students learn under the tutelage of teachers and become experts, in vernacular knowledge, expertise is developed through the logic of practice and outside the four corners of an educational system. That is, knowledge is contingent; it emerges to solve situational crises. Since vernacular user knowledge is contingent it could be characterized as situational, local and context dependent. Consequently, knowledge associated with the latter forms is devalued because they do not fall under what can be considered as "expert knowledge."

It is thus not by mere happenstance that the writers of this report classified the measures used as "unscientific." However, for me, this is an attempt to devalue the knowledge that emerged to solve the problem. We are also reminded by Robert Johnson that "within our knowledge of the mundane, we often act and do as specialists, but we are not allowed to claim such knowledge because (in most cases) we were not taught such knowledge in a formal educational environment" (p. 6). This extract by Johnson, perhaps, explains why writers will label the heuristic measures adopted as "unscientific:" the measures they adopted were not learned in the classroom thus they cannot be classified as "scientific" or "fine knowledge." Since it was a voting situation, it could even be that most of the users who devised and used these strategies have never stepped foot in the classroom, the more reason why their knowledge is devalued. Society, by sheer use of categorization and by means of stratification, has managed to place much authority on the educated, a domain fairly associated with elites or literacy over the uneducated, not elite or popularly known as illiterates. Thus, in the scenario above we can even choose to replace "unscientific" method with illiterate or vernacular method. This last line explains perfectly what it means to devalue the knowledge that emerges from an illiterate-not-so-refined-mind. It is therefore not wrong to say that the categorization of users as knowledgeless is not limited to Western societies, but it is a phenomenon that permeates societies.

Regardless of the categorization, I consider this move to use Coca-Cola, OMO detergent and other simple means to resolve issues and get voters to vote a problem-solving activity. It is also a confirmation of Sun's words which opens this chapter. The users were able to make sense of the situation and creatively customized and localized a technology to fit into the electoral system, contributing to their identity-building process. Thus, localization should not solely be about how designers change technological features to meet local situations; users also influence design decisions at micro-levels. We do not pay attention to user knowledge because of the perception that user knowledge is spontaneous and informal.

I refer to the use of unscientific and vernacular methods to resolve breakdowns with technology as user-heuristic experience localization. At the center of user-heuristic experience localization is the idea that user knowledge is not regulated by formal rules and procedures learned in the formal education system or through any formal means; instead, users develop their expertise and form of problem-solving in response to unpredictable real-life situations that arise in regard to technology use. I emphasize innovative user strategies to resolve the mass breakdown of the biometric technologies used during the 2012 elections. I also emphasize how the local, vernacular solutions coupled with broader systemic thinking about the 2012 experiences have helped Ghana to properly integrate the biometric technology. I examine the various moments when users drew upon and created what literacy scholars such as Barton and Hamilton (1998) refer to as "local knowledge or vernacular knowledge" (p. 242) to solve the problem.

My discussion of user-heuristic experience is articulated most clearly by Johnson-Eilola's and Stuart Selber's inundation of heuristic problem-solving approaches. In *Solving Problems in Technical communication*, the scholars discussed how heuristics provides "rough frameworks for approaching specific types of situations...not by providing straightforward answers but by providing tentatively structured procedures for understanding and acting in complex situations" (p. 4). They believe that "problem solvers possess several important characteristics, including the ability to sense a problem, diagnose what forces within a context are causing the problem, and develop and implement a change within the context that addresses the problem" (pp. 3–4). This is an indication that users' fall on resources available to them to solve complex technological problems. Heuristic frameworks as I will indicate in this chapter involve stepping back, becoming aware of the situation and adopting "different approaches based on intuition (itself based on experience)..." (p. 5).

It is without a shred of doubt that the handlers of the biometric technology, the EC officials, and voters exhibited most of the traits Johnson-Eilola and Selber discuss as heuristic problem-solving. The EC officials became aware that the weather could not support the biometric technology, so they had to figure a way out; so did biometric handlers and voters decide to use detergents such as OMO and even Coca-Cola to clean the hands of those rejected. Reports indicate that these ad hoc measures proved useful. For instance, Gadugah narrates that "I remember one community in my district, none of the people could be verified... we were told that if the people washed their hands with Coke, it would work and we started and as soon as we started washing their hands with Coke it started picking up." The examples indicate that heuristics can be user-based, or user-generated but still be theoretically grounded. Users work with theory, though implicitly, to solve problems and it also points to the fact that there is a theory to practice. This means that users have a form of cunning knowledge (a concept I will dwell on in the next chapter), although it is disregarded, that helps them to conceptualize and act at the opportune time. These cunning forms of knowing and acting includes stepping back, learning from experience and sense making.

I believe that because of the positive representation of the biometric device, EC officials did not put in place mechanisms that will prevent breakdowns or they simply ignored warning signs because they felt the machines were foolproof. In my introduction, I stated how Mr. Gadugah found it funny that trainers kept mentioning that they should avoid contact with sunlight or dust because no one in their right senses will carry the machine and go and sit in the sun, so "So what exactly do you mean by we shouldn't be in direct sunshine, okay?" he quizzes, and Mr. Manu also said the Commission thought that they could use canopies as shades for the biometric. This ad hoc measure was criticized vehemently by Mr. Gadugah who told me that "...if you ever sat under these canopies, the heat that the canopies radiate is even more than going directly under the sunshine." It is therefore important to interrogate crisis moments that surround the deployment of biometric technology. In focusing on biometric failures and what it teaches us about technology failures, I emphasize the significant role users played in saving the technology. User activities during the breakdown of the biometric technology at polling stations confirm my argument that users are heroes who do not wait on developers.

ACTIVE SENSE-MAKING DURING TECHNOLOGY USE

In *Cross-Cultural Technology Design*, Huatong Sun emphasized the relevance of sense-making to localization processes. In Sun's estimation, "when a user is able to make sense of his or her particular technology use, we say that the user experience has come into being" (p. 218). This is an indication that a proper user experience is achieved when a user is able to work to construct meaning around technology, or that the user should work to fully understand the technology and ask whether or not that technology is able to support local activities. Karl Weick, the proponent of "sense-making," defines the term as simply "the making of sense" (p. 4); and Barton and Hamilton (1998) identify that sense-making "is motivated by the search for meaning and problem solving" (p. 231). This means that sense-making has to do with figuring out or "deliberating on" in order to understand the unknown or solve an emergent problem. Sense-making provides heuristics as this process of meaning-making does not follow a specific logic. It goes back to Johnson-Eilola and Selber's characterization of heuristics as a rough framework. Therefore, sense-making and heuristics are used as problem-solving tools since both concepts seek to provide an understanding of an underlying problem.

In *The Design of Everyday Things*, Don Norman identified that good designs have two characteristics: discoverability and understanding. Discoverability enables people to figure out what actions the technology can perform and how to perform them; while understanding is sought when users ask how everything that comes with technology works: "what does it all mean?" (p. 3). In addition, Norman identifies that users of technology fall into two gulfs: the gulf of execution and the gulf of evaluation. In the former, the user is figuring out how things work with a technology and in the latter, users evaluate or reflect on what happened when the technology was put to use (p. 38). The first of the gulfs, execution, is related to sense-making approaches. Thus, if one is able to discover and understand the characteristics of the technology, that user uses the device effortlessly and pleasurable. These two concepts as explained by Norman are summed up perfectly in Weick's sense-making. Making sense of technology use, thus, has to do with how users work to make meaning or construct meaning around the technology in use. It involves constant questioning of the systems to find an answer to how that system works. Meaning-making around technology is very necessary because neglect of the various meanings users make could lead to potential breakdowns. When machines fail, little do we blame it on designers, we chastise the user "for not understanding the machine" (Norman, 2013, p. 6).

The conversations I had with Gadugah and Manu above, bring into reality the relevance of sense-making to localization processes. We encounter efforts made by these EC officials to situate the technology in a broader Ghanaian context or efforts made to understand the logic of the biometric technology and get it to work in the Ghanaian context. Of particular concern here is whether weather will support the technology use or what methods could be applied to make the biometric function properly considering that the technology was not developed to withstand the harsh weather conditions of the country. Gadugah, for instance, did not stop telling me that the technology was not "robust enough." We even encounter how an ad hoc solution to use canopies to shield biometric technologies from the scorching sun introduced new problems for the Electoral Commission of Ghana as the canopies got heated up and consequently escalated the breakdowns.

Stepping Back as a User-Heuristic Strategy

Ghana's use of the biometric technology indicates that stepping back, learning, and experience are central to user-heuristic localization approaches—I also emphasize that these are rhetorical strategies users of technology employ when they find themselves in critical situations.

Users do not get to acquire these traits easily, through educational lectures or reading, but laboriously through past mistakes/encounters. At the level of user-heuristic experience, user moves away from linguistic prestidigitations to practical forms of knowledge making; they think about broad contextual use situations of a technology than a mere use of flamboyant language to describe the functional attributes of a technology (linguistic localization). To a large extent, stepping back, learning, and experience are akin to Johnson-Eilola's "system thinking," Banks' "critical access," and Seigel's "disruptive-system documentation." In "Relocating the Value of Work," Johnson-Eilola suggests among other things that technical communicators should be trained to be system thinkers, that is, they should learn to be able to "recognize and construct relationships and connections" (p. 261) in ways that move them beyond traditional problem-solving approaches. For Johnson-Eilola, system thinkers work to "understand (and remake) systemic conditions." This move will ensure that technical communicators do not merely break down problems into manageable chunks to be solved in small bits; but instead, they will "step back to look at larger issues in the system to determine how the problem develops and in what contexts it is considered a problem" (p. 261). This form of thinking proposed by the author is exemplified in Banks (2006) notion of critical access which places much emphasis on the need of the user to develop and understand both the good and bad ends of a technological system before they make a commitment to use or resist it (p. 42); and Seigel's (2013) call for system disruption tendencies which helps users to "manipulate parts of a system, negotiate the system, or change the system even in a small, local way" (p. 74).

I cannot say with certainty that Ghanaian users of the biometric worked to manipulate parts of the biometric system, except for a presiding officer who wrote 270 in words as "twenty seven zero (Baneseh, 2015, p. 85)," but it is obvious they worked to negotiate the system to get it to work to benefit the electoral process. Earlier, I mentioned how EC officials tried to control climate conditions surrounding the biometric technology by providing canopies to serve as shades. I also talked about how those rejected were caused to use Coca-Cola and some other detergents to wash their hands during the voting process. Subsequently, after the 2012 elections, the Electoral Commission and other stakeholders adopted a systemic thinking approach when they reflected the role of the biometric technology in the electoral process of the country.

This tendency to step back to look at the broader issues in the system to think about how to resolve problems became very helpful as subsequent elections registered less breakdowns. For instance, a preliminary report on the 2016 elections released by the CODEO captured only 26 breakdowns (p. 8) across the country. When asked what the commission did, Kwaku Manu stated,

> …so the first point is that we had a look at all these things: why some of the BVDs actually failed; why some pink sheets were not properly filled; why maybe there were some over voting and other things in some polling stations and other things…Then we also took a look at why the BVDs were frequently breaking down and we saw that some of them did not handle them very well, some of them didn't go by the operational instructions some of them weather conditions and all those things so the commission said that okay then why is it that we don't put two BVDs at each polling station and get some back up also so that if there is a problem, quickly the second one can be used and that is the reason why we did not hear about breakdown of BVDs because once one is faulty then the other one is replaced immediately and then used [you know] so if you look at ehmm these two issues, you find out that these are some of the things the commission actually did.

Learning from Experience as User Strategy

Mr. Gadugah also asserted that:

> …So learning from the challenges of 2012, when we went into 2015 there was an improvement, okay? Then we used the BVDs for 2015 so we improved on the efficiency and effectiveness then come 2016 we provided enough. Every polling station had 2 BVDs, backups at the district and backups at the regional offices. So where there are breakdowns, immediately you could bring in the backups. If, in the unlikely event that the two backups broke down on Election Day, you could also get an immediate backup from the district or the region. So learning from the shortcomings of 2012 and the fact that we again used the same device from the revision exercise in 2014 and the election in 2015 it informed the training, the deployment and the quantity of BVDs that we used in 2016.

The extracts above indicate that though not much was done to change the biometric technology, several forms of negotiations went on after the 2012 elections. By stepping back to look at the issues, the EC officials were exhibiting the form of thinking that Johnson-Eilola advocated in his article, the system disruption attitude Seigel advanced, and critical thinking. The experiences above indicate that experimentation is not the only means to arrive at knowledge; intuition and experience move users to act in productive and knowledgeable ways. In fact, it can even be adduced that this heuristic strategy of stepping back, and learning from experience follow a methodological approach: (1) the user encounters a problem in a technological system; (2) user steps back to make sense of the problem; (3) the user intuitively comes out with a solution; (4) the solution is tested; and (5) the user waits to find out what the result is. If this heuristic approach fails to work, the user quickly steps back and tries out other meaningful strategies. It is without a doubt that we teach critical thinking in college, and yet these simple, uneducated people seem to have exhibited understandings and abilities that are still seriously lacking in college-educated people. This is an indication that users have forms of knowledge that can help educated, trained technology designers.

Users as Critical Thinkers

User-heuristic localization captures users as reflective individuals. As I said earlier, users ask broader questions about the system to get a broader picture of the role of a technology in a system. In this section, I focus on Honorable Dr. Akoto Osei's contribution to parliamentary deliberations on the biometric technology and how he helps us to understand what it means to be critical of technology; a standpoint my research advocates. Of all the conversations captured in the report, only Dr. Akoto Osei shifted the stasis from procedural one to evaluative: How do we critically ensure that the biometric we imported would work accurately? He shifted the argument from procedural to evaluative based on his experience with the technology.

I would like to point out that the question posed by Akoto Osei: What happens when biometric fails to operate on the Election Day provides a framework to discuss what I refer to as a critical engagement or Banks' idea of "critical access" to technology. According to Banks (2006), it is not enough to provide a technology to users; instead, it is paramount for individuals in a particular community to develop an understanding of "the benefits and problems of any technology well enough to be able to critique, resist, and avoid them when necessary as well as using them when necessary" (p. 42). Hence, through Akoto

Osei's narrative, we can defuse the notion that biometric technologies are error free. In fact, they are not error free and they are subject to breakdowns. Like the breakdown and failure narratives presented by Akoto Osei and the Deputy Speaker of Parliament, Magnet (2011), identifies that the biometric system is subject to three errors: false acceptance rate, false rejection rate, and failure to enroll. False acceptance rate occurs when a person who is not you is accepted as you; false rejection occurs when you are not accepted as you; the failure to enroll error occurs when the system fails to enroll someone (p. 22). As the failures above indicate, the biometric technology did not always function in a straightforward way as purported in the documents I have analyzed. I conjecture that this is the kind of breakdowns we encounter when we ascribe to the notion that technologies are foolproof and cannot breakdown. When we submit to the internal logic of technologies, we risk being critical of the technologies. More importantly, we risk asking questions that will help to ensure the successful use of technology. In essence, when we submit to internal logic of such technologies as the biometric, we fail to "examine critically the nature and significance of artificial aids to human activity" (Winner, 1986, p. 4). These stories by politicians and election observers destabilize our notion of the biometric as a technology that is not subject to failure; the biometric is not a perfect tool after all.

Ghana's experiences with the biometric indicate that users can adopt strategies and tactics that enable them to think broadly about the system within which the technology is being integrated. In this process, users step back, learn from experiences, and make sense of what strategy to adopt in a situation. These are indicative of the fact that the expertise users bring to technology use can be extensive and has many dimensions. My emphasis on user creativity re-echoes earlier attempts by scholars to stress the agency of users by focusing on how they creatively customize and localize technology to fit their local situations. Huatong Sun, as an example, contends that such creative impetus from users contributes to their identity building. Indeed, such is what I seek to do in this chapter, to look at the creative ways users approached the biometric breakdowns that almost marred the beauty of the 2012 elections. My goal is to stress the fact that users do not wait for designers to come to save them, they know how to save themselves. The actions of users, namely the EC officials, community members, and voters, reemphasize my earlier claim in Chapter 5 that users embody specific user knowledge which enables them to judge the moment and adopt appropriate solutions.

Therefore, user-heuristic, like the other localization processes I have discussed, emphasizes the agency of users, but this kind of localization has an added characteristic of "learning from mistakes."

References

Arhin, A. (2016). *CODEO'S interim statement: Observation of the on-going voter register exhibition exercise.* Retrieved from Accra, Ghana.

Baneseh, M. A. (2015). *Pink sheet: The story of Ghana's presidential election as told in the Daily Graphic.* Accra: G-Pak Limited.

Banks, A. J. (2006). *Race, rhetoric, and technology: Searching for higher ground.* Mahwah, NJ and Urbana, IL: Lawrence Erlbaum and National Council of Teachers of English.

Barton, D., & Hamilton, M. (1998). *Local literacies: Reading and writing in one community.* London: Psychology Press.

Final Report on Ghana's 2012 Presidential and Parliamentary Elections. (2013). Retrieved from Accra, Ghana. http://www.codeoghana.org/assets/downloadables/Final%20Report%20on%20Ghana's%202012%20Presidential%20and%20Parliamentary%20Elections.pdf.

Hobbis, S. K., & Hobbis, G. (2017). *Voter integrity, trust and the promise of digital technologies: Biometric voter registration in Solomon Islands.* Paper presented at the Anthropological Forum.

Johnson, R. (1998). *User centered technology: A rhetorical theory for computers and other mundane artifacts.* Albany: State University of New York.

Magnet, S. (2011). *When biometrics fail: Gender, race, and the technology of identity.* Durham, NC: Duke University Press.

Norman, D. (2013). *The design of everyday things: Revised and expanded edition.* New York: Basic Books.

Norman, D., & Nielsen, J. (2008). *The definition of user experience.* Retrieved from https://www.nngroup.com/articles/definition-user-experience/.

Seigel, M. (2013). *The rhetoric of pregnancy.* Chicago: University of Chicago Press.

Sun, H. (2009). Designing for a dialogic view of interpretation in cross-cultural IT design. In N. Aykin (Ed.), *Internationalization, design and global development* (pp. 108–116). Heidelberg: Springer.

Sun, H. (2012). *Cross-cultural technology design: Creating culture-sensitive technology for local users.* New York: Oxford University Press.

User Experience Network. (2008). Retrieved from http://www.uxnet.org/.

Winner, L. (1986). *The whale and the reactor: A search for limits in an age of high technology (La Baleine et le réacteur).* Chicago: Chicago University Press.

CHAPTER 7

Subversive Localization

> We realized that the user manual that came with the biometric did not say anything about how to use the biometric device to register people for elections and the illustrations were not enough to help the users understand the processes involved in operating the system so we had to redesign it [user manual] and relate the use of the machine to the electoral process of Ghana —Mr. Agyapong

Localization is subversive, by which I mean, it challenges dominant ideas that cast technology designers as experts and users as dummies. In this chapter, I focus on the user manual which accompanied the biometric technology adopted by Ghana to reinvigorate my argument that users are not dummies but heroes who reconfigure, shape, redesign, and localize an available technology to fit into their local contexts. It is important to state from the onset that the manual was not designed by the designers of the biometric technology; instead, it was designed by EC officials who were far removed from the biometric design. I must also add that it is possible that none of the EC officials who were part of the manual design had any training in instruction design or technical writing. How else could they have done that? What does this process of redesigning the user manual by users mean to the technical communication pedagogy and research in an era of technology flows? The actions of the EC officials speak to one thing: we need to study localization from bottom-up. That is, we need to value the knowledge users bring to localization process. This process of redesigning to meet local needs provides

© The Author(s) 2020
I. K. Dorpenyo, *User Localization Strategies in the Face of Technological Breakdown*, https://doi.org/10.1007/978-3-030-26399-7_7

a lens through which I can analyze and articulate one major localization process: subversion or reconfiguration.

By focusing on the manual, I indicate that while users can use heuristic approaches to resolve technological breakdowns, as discussed in the previous chapter, they can also subvert official knowledge by dwelling on their cunning intelligence in their bid to solve complex technological problems. I call the determination of users to redesign technologies or documents to fit local contingency *subversive localization*. I use subversive localization to refer to instances when a user picks "up available tools, adapts them in idiosyncratic ways, and makes do" (Spinuzzi, 2003, p. 2). They achieve this by creating "innovations" that help them to subvert the available documents or technology, invent their own ways to turn them to their needs (Spinuzzi, 2003, p. 2). To be clear, "subversion" as I use it in this book refers to the extent to which users reconfigure the original intent of a technology. As I have indicated throughout this book, "reconfiguration" is an important part of localization processes. While literature is replete with examples about how designers of technology reconfigure technologies to meet different locales, little is done to study how users reconfigure technologies they have adopted and made them meaningful to their context. To this end, users have been cast as knowledgeless. The extract above from Mr. Agyapong is a tacit confirmation of the knowledge that users wield: users know their context, they know what works for them and what does not work for them. The Electoral Commission officials, as an example, realized that the original manual which came with the biometric did not capture context of use, that is, it never educated users on how to use the biometric to conduct elections. To affirm their knowledge and expertise, they design their own manual. It is important to note that none of the EC officials has had any training on instruction writing, and I doubt any of them has had any training in technical communication because this field has not been fully developed in Ghana.

My argument in this chapter is not different from those I made in preceding chapters: Users embody some kind of "practical wisdom," that is, "a true and reasoned state of capacity to act with regard to the things that are good or bad for man" (McKeon, 2009, p. 1026). Aristotle makes us understand that a person with practical wisdom is "able to deliberate well about what is good and expedient" (McKeon, 2009, p. 1026). Thus, the capacity for user to act, I believe enables them to judge the appropriate moment to revise or redesign documents imposed on them. Therefore, if users have the capacity to act, as I will indicate they do, then they cannot be portrayed as victims who are waiting to be rescued. They are able to initiate their innovative or creative strategies which help them to accomplish their goals.

The manual does not only provide instructions for how to use the biometric device, but it also became a legal document used as evidence to support an argument at the Supreme Court of Ghana. As I indicated in Chapter 2, after the 2012 elections, the opposition NPP filed a petition contesting the credibility of election results. During this petition, the user manual became a legal document which was tendered in as evidence to reveal inconsistencies in election laws guiding the 2012 elections. For instance, while it was mandated that people who were rejected by the biometric were not supposed to vote, pages 16 and 20 of the user manual indicates otherwise: "if the biometric verification is not successful, the presiding officer will be alerted and he/she will take a final decision on the voter" (manual, p. 16). In fact, page 1 of the manual states that "a voter whose fingerprints are not recognized by the device will not be allowed to vote." The counsel of the opposition NPP interpreted this inconsistency to mean that the EC deliberately trained presiding officers to allow people to vote even if they had been rejected. He also interpreted it to mean EC gave power to presiding and verification officers to disobey election laws (Baneseh, 2015, p. 122) even though the EC argued otherwise. It became apparent throughout the exchanges between Mr. Addison, counsel for petitioners (NPP), and Dr. Afari-Gyan, the Chairman of the EC, that the user manual did not only contain guidelines on how to operate the biometric device, it also contained legal codes that controlled the behavior of EC officials at the various polling stations. More so, the EC officials I interviewed indicated that some verification officers could not follow the instructions laid out in the user manual, hence the breakdown of biometric devices in some polling stations. This last point indicates that failure to follow instructions does not only lead to serious injuries but could potentially disenfranchise voters or considering that elections are high stakes, it could even plunge an entire country into chaos.

James Paradis (2004) has indicated that technical documentation is "exegetical," meaning it serves to make meanings, applications, and procedures simple for lay people who use technologies to perform certain tasks (p. 366). In other words, it is the bridge between experts of technology design and users. He reveals that operator's manuals, for instance, provide avenues for "written discourse to construct a world of 'reasonable' actions that would resolve the polarities inherent in almost any technology between mechanical function and social purpose" (Paradis, 2004, p. 367). In this role of mediation, operators' manuals employ four textual elements that bind human action and external objects with human behavior. They (1) construct a written analog of the tool or process itself; (2) introduce a fictional operator who represents an average or suitably qualified individual; (3) capture "the material context of conditions and situations requisite for effective and safe use of the instrument" (p. 387); and (4) capture the action itself. In this sense, the operator's manual fuses human purpose into a mechanical device. Instructions mediate the gap between experts of technology and lay users. In transporting technologies to the lay public, instructions play "a necessary role in the exportations of expertise, whether simple consumer technologies or complex instruments of institutional proportions" (p. 371).

This role of written discourse gives texts permanence in the world. Users can learn how to operate an artifact by operating on the text. One can also not dispute the legal ramifications that operator's manuals carry along. Legal ramifications point to the fact that writers are held accountable to what they scribble. The operator's manual presents two affordances: (a) we are persuaded into believing that the technology can be explained; and (b) it is the evidence that the expert has reasoned out about how the user can use the technology. It is a warranty. Ghana's case, for example, exonerates Paradis' argument that documentation, such as a user manual, does more than merely guiding users on how to use a specific technology.

Regardless, of the relevance of manual as a legal document, the primary focus of scholarship about documentation seeks to focus on "what makes instructions effective and ineffective" (Seigel, 2013, p. 8). This means that researchers have been concerned about how technical communication is able to help users understand a specific technology and use it. It is probable to say that in this line of inquiry, researchers or designers are only interested in collected data on users so they can improve on communications about the technology to their users. I have indicated in Chapter 5 that when we focus more on designing documentation to accommodate technology to users, we are more influenced to emphasize the instrumental features of that technology we are communicating about and not think critically about the role of the technology in a broader system. More worryingly, we do not focus on user strategies adopted to get technology working when instructions are not effective. In this regard, we only pay lip service to claims that we are user advocates. We do not advocate for users if we are not so interested in their creative strategies. I contend that when instructions do not work, users make local moves to help make technology successful so we must pay attention to such innovations at local contexts. When this is done, we will acknowledge that users embody a specific kind of knowledge.

Other reasons account for my interest in the user manual: Documentation has long been the province of technical communication scholarship (Johnson, 1998; Paradis, 2004; Seigel, 2013). Scholars have proven that "written instructions govern, guide, and control user actions on a daily basis in tasks that range from operating industrial equipment, installing a wireless router, to using computer software" (Hogan, 2013, p. 155); documentation can "sustain" a system or disrupt a system. "It is the rhetoric that accompanies technology and makes it possible—the rhetoric that makes technology fit into the world and makes the world fit with technology" (Bazerman, 1998, p. 388), and it is "one of the things that determines how we use technology and in turn how technology uses us" (Seigel, 2006, p. 13). Moreover, Ghana's increasing economic development and business exchange with the West, especially the US, make it possible that the Ghanaian will purchase more items from America or the West. Hence, it is possible that technical communicators in the West will write documents for Ghanaian users. It is also possible that Ghanaians may collaborate to write documents for Ghanaians or Western audiences.

The Manual as an Articulation of Instruction Writing 'Pieties'

The manual, as I indicated in Chapter 5, is a 31-page booklet produced by the Training Department of the Electoral Commission and SuperLock Technologies Limited (STL), the providers of the biometric device. The manual serves as a training guide and a reference tool for operational staff, notably key trainers and polling staff who will handle the biometric device at the various polling stations. It mostly contains the procedures and legal precedents that articulate the operation of the biometric device. Words and phrases used are meant to identify the different aspects of the process in using the biometric device and what polling agents should do when they encounter such issues as the inability of biometrics to identify voters.

I describe the features of the instruction manual in light of the guidelines to writing manuals that are taught in traditional technical communication classrooms in the US. My decision is based on the fact that these guidelines are what technical communicators or designers "typically consider when planning, drafting, and revising" (qtd. in Dong, 2007, p. 224) instruction manuals. In a way, the guidelines outlined in the books I have selected can be considered the best practices. Seigel (2013) refers to these best practices as "pieties" that users or in this case, writers must follow. I review four technical communication textbooks I have used at one point or the other in the ENGH 388: Professional and Technical Writing classes I have been teaching at George Mason University and HU 3120: Professional and Technical Communication classes I taught at Michigan Technological University when I was a graduate student. A critical study of these four technical communication textbooks used in my introductory technical communication classes in the US: *Writing that Works*, by Walter Oliu et al.; *Strategies for Technical Communication in the Workplace* by Laura Gurak and John Lannon; *Practical Strategies for Technical Communication* by Mike Markel; and *Technical Communication: A Practical Approach* by William S. Pfeiffer and Kaye A. Adkins outline some best practices that instruction manuals must follow.

All four textbooks indicate, for instance, that it is necessary for an instruction manual to have an introduction or an overview. They hold that the introduction or the overview works to explain the purpose of the instruction or all of the required information the user needs in order to perform the instructions (Gurak & Lannon, 2012; Markel, 2013; Pfeiffer, 2003). The best practices also reveal that an instructional manual must: describe a sequence of steps; group items under task headings and subheadings; and be written from a command point of view using an imperative voice. That is, each step must proceed with an action verb; speak to the rhetorical situation; use accessible language; emphasize caution, warning, and dangers; use visual symbols and signal words; and contain a title, brief overview/general introduction, step-by-step instructions, and conclusion. Instructions must follow these guidelines in order to enable users to perform their activities or actions (Gurak & Lannon, 2012; Oliu, 1994; Pfeiffer, 2003); or "they must get out of the way of the user" (Seigel, 2013) so they can operate a technology. One of the four books I reviewed suggests that clearly written instructions should "enable your audience to carry out the procedure/task successfully" (Oliu, 1994, p. 415).

The manual fulfills most of the elements that an introduction to a user manual must exhibit. The manual begins with a "Foreword" which was written and signed by the Chairman of the Electoral Commission, Dr. Kwadwo Afari-Gyan. The "Foreword" serves as an introduction to the manual and it gives an insight into why the manual was produced, to whom it was produced and why the biometric system was adopted. It basically identifies the rhetorical situation. For instance, it is written in the foreword that the manual "contains procedures for operating the biometric verification device" and it will "serve as a training document and a reference tool for the operational staff" ("The Biometric Verification Device User Manual," 2012, p. 1). Also, the foreword introduces the importance or purpose of the device: It was "introduced to the electioneering process in Ghana as a means of enhancing the integrity of the electoral process of the country" and it was going to "assist in the detection and prevention of practices of impersonation and multiple voting that have characterized our electoral process in the past" (ibid.). Subsequently, the foreword introduces the audience of the device as "key trainers and polling staff who will handle the device at the polling station, as well as key stakeholders." These "trained personnel" will use it to "authenticate the identity of voters by means of finger prints" (ibid.). In addition, the rule that guides the use of the biometric system is revealed: "a voter whose fingerprints are not recognized by the device will not be allowed to vote: while voters whose fingerprints are recognized by the device will be issued with ballot papers and allowed to cast the votes at the polling station" (ibid.).

What the foreword to the manual accomplishes is a barrage of promises about the successes that will be chalked when the technology is used. Little room is put into commenting on the quality of the biometric device. In this sense, the manual becomes an example of an epideictic documentation. By "epideictic" documentation, I mean to say that the manual is only "concerned with present action" (Seigel, 2013, p. 71). It provides "functional but not conceptual aspects" (qtd. in Seigel, 2013) of the biometric device. The sole purpose is to get users engaged with the biometric so that results are perceivably captured accurately and efficiently.

Also, these textbooks used in technical communication classes reveal that it is *sine qua non* to use headings and subheadings when writing instructions. The purpose of grouping tasks under headings and subheadings, is twofold: First, "they divide fragmented information into manageable chunks" (Pfeiffer, 2003, p. 225) for readers. Second, "they give readers a sense of accomplishment as they complete each task..." (ibid.). The manual meets this criterion as sections are organized under main headings which are captured in caps, and subheadings captured in lower cases. Headings such as "the biometric verification device," "how to unpack the BVD," "tips for using the BVD," and subheadings as "functional elements of the BVD," "sleep mode of the BVD," and "how to replace batteries" help readers to move to necessary sections of the manual in order to use the technology efficiently.

Some other peculiar features of the manual need to be mentioned. The manual relies almost entirely on visual cues such as headings, labeling, pictures of the various stages of registering, images of messages to expect at each stage of the process, and warnings. For example, in Fig. 7.1, users are introduced to the functional parts of the biometric device.

I would like to point out, however, that while some command prompts do not reflect their headings, some other visuals are placed haphazardly. For example, Fig. 7.2 has the heading: "how to unpack the BVD," but steps under this section say nothing about how to unpack the device.

A similar case is found in the section headed "Message requiring cleaning the fingerprint scanner" in Fig. 7.3.

Some images are also placed in document without any accompanying texts to indicate what they mean. Figures 7.4 and 7.5 are tacit examples.

Also, best practices of instruction writing state that the body of an instruction manual must be written in steps. Steps should begin with active verbs and they should be phrased as commands (Gurak & Lannon, 2012; Markel, 2013; Oliu, 1994; Pfeiffer, 2003). According to Markel (2013), steps form "the heart of a set of instructions" (p. 381). The importance and rationale in phrasing steps as commands or tasks is grounded in the idea that "users want to spend a minimum amount of time reading instructions" (Seigel, 2013, p. 78). Users want to use their products to achieve a goal. Seigel states that this advice is because we assume that instructions are to maintain systems and not disrupt them. As I stated earlier, the user is made to engage with the system through the tasks they perform. In this regard, one only focuses on the promises the system provides: You can only achieve your goal if you use it (the

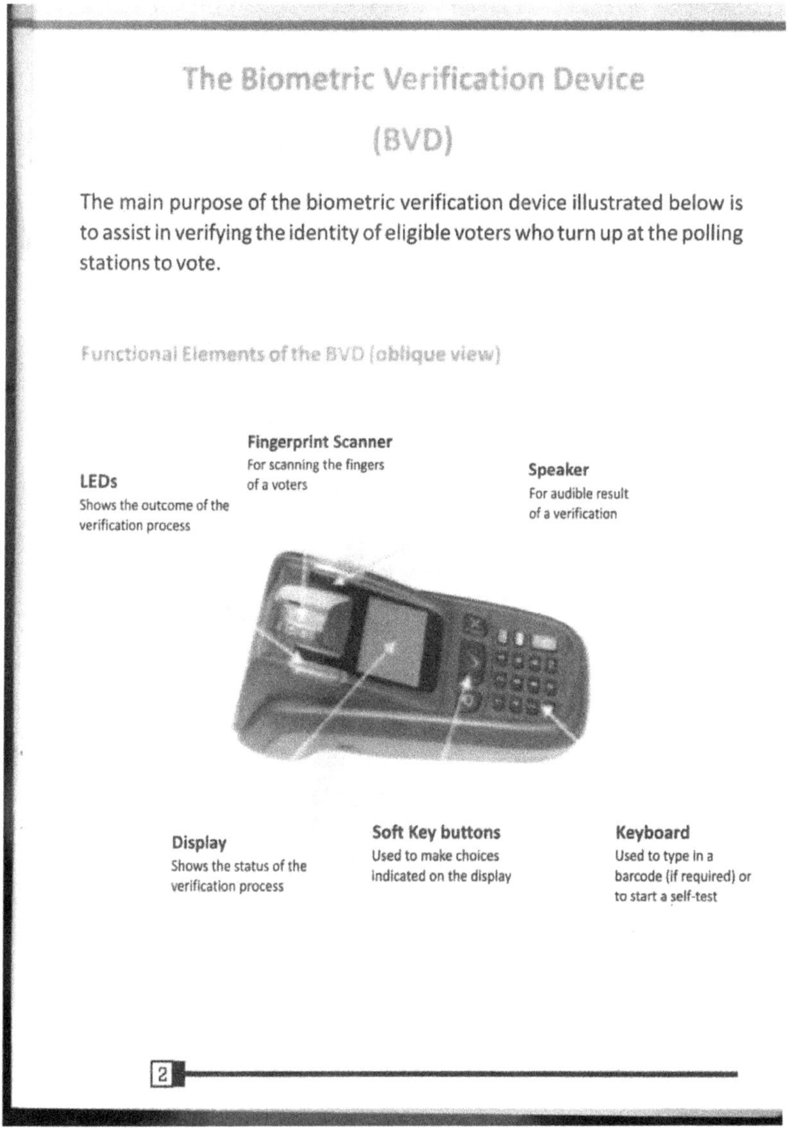

The Biometric Verification Device (BVD)

The main purpose of the biometric verification device illustrated below is to assist in verifying the identity of eligible voters who turn up at the polling stations to vote.

Functional Elements of the BVD (oblique view)

LEDs
Shows the outcome of the verification process

Fingerprint Scanner
For scanning the fingers of a voters

Speaker
For audible result of a verification

Display
Shows the status of the verification process

Soft Key buttons
Used to make choices indicated on the display

Keyboard
Used to type in a barcode (if required) or to start a self-test

Fig. 7.1 Functional elements of the biometric device

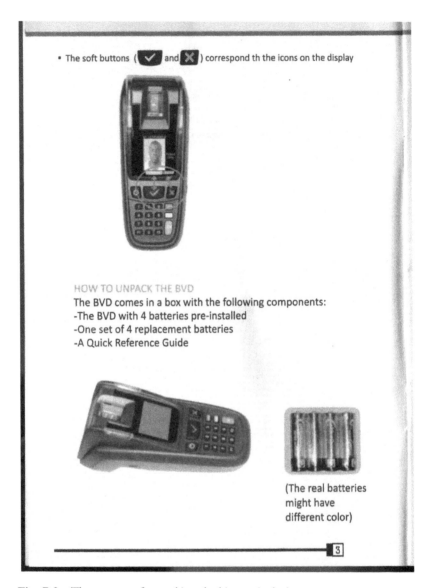

Fig. 7.2 The process of unpacking the biometric device

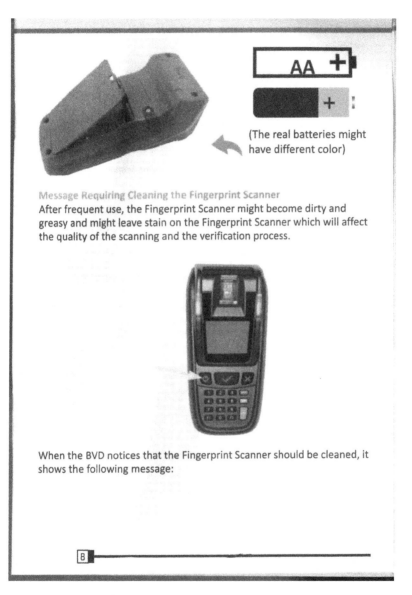

Fig. 7.3 Message indicating when to clean the scanner

Maintenance Messages of the BVD
The BVD internally checks its own status and issues messages when there is the need to perform maintenance.
Other messages that will be displayed by the BVD and the required actions to be taken are indicated below.

LOW BATTERY POWER
- When the battery power is getting too low, the BVD will indicate by showing the following message:

- The required action is to *Replace the Batteries*

Fig. 7.4 Message indicating biometric should be handled with care

7 SUBVERSIVE LOCALIZATION   157

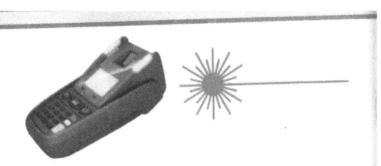

b. Scanning Finger-print
After Scanning the Barcode, the next process of identifying the voter is to scan the fingerprint.
When scanning a fingerprint using the Fingerprint Scanner, do the following;
- Place the BVD on a *steady surface* (e.g. a table)
- Make sure that the surface of the table is *smooth* and that the BVD is *stable* (not wobbly)

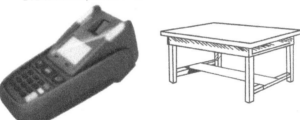

- The left index finger is first placed flat and *in the middle* of the Fingerprint screen
- The finger is placed firmly (but not too strong)
- The finger is kept on *the scanner* until it is captured. The device will request the voter to place a particular finger if it does not recognise the first one. There is the possibility that a number of fingers may have to be placed on the screen one by one until it picks a valid fingerprint.

Fig. 7.5 Laser beam

system). In the case of the user manual that I analyzed, we can say that users are instructed to focus on tasks because they have to live to the promises that the biometric system provides: to assist in the detection of malpractices. Thus, an electoral process that has been historically, socially, politically, culturally, and legally marred by deep-seated malpractices can be corrected if users piously followed laid down procedures about the system to the letter.

Not only are instructions supposed to be task-oriented, they must also be written in imperative voice (Gurak & Lannon, 2012; Markel, 2013; Oliu, 1994; Pfeiffer, 2003). We are encouraged to use imperative voice because imperatives clarify "when the user her-or himself should be carrying out a particular action" or "instructions carry the weight of expertise, of authority" (Seigel, 2013, p. 82). In essence, instructions represent the system experts. These suggestions, Seigel indicates, carry implicit political and ideological undertones. For instance, although *The Biometric Device User Manual* does not explicitly say that if one fails to follow instructions to the letter the promise of delivering credible results will be compromised; this argument is implicitly implied in discourses surrounding the biometric technology. It is argued, for example, that the biometric device internally checks its own status and issues messages when necessary; or that the BVD gives an indication to the user when battery power is getting low. Relatedly, when BVD notices that the fingerprint scanner should be cleaned, it indicates. When such messages pop up, the user only has to either "replace the batteries," "clean the fingerprint scanner," or take an action when necessary. These discourses implicitly push forward the promise that the device will enhance the integrity of the electoral process of the country by assisting to detect or prevent malpractices. The implication is that the system is foolproof; hence, any user of the technology can operate the BVD because the machine has in-built security mechanisms.

The idea that anybody or any user can operate the BVD successfully received little resistance from the populace as discourses persuaded Ghanaians to trust and have confidence in the system. This promise of credible electoral results was, however, challenged when after the elections the NPP petitioned the Supreme Court over malpractices that occurred during the election. More so, after the election, we encountered such newspaper headlines as "EC 'Steals 1.3m Votes for John Mahama;'" "More Issues Over Election Results;" "Let's Recount Votes;" "Making Nonsense of Election Technology," etc. These headlines critically looked at not only the BVD but the users of the BVD as well. Thus the promise made by the Chairman of the Electoral Commission is subjected to scrutiny; ideologies about BVD contested.

In *The Biometric Device User Manual*, instructions include prescriptions about how to scan fingers of voters; how to unpack the biometric system; how to place fingers of voters on biometric device; how to switch device on/off; how to respond to maintenance prompts; how to scan barcodes, etc. Though some of the tasks are phrased in commands, most of the tasks are not phrased as such. Instead, instructions are presented as descriptions. For instance, in Fig. 7.6, under "how to switch on/off the BVD," the command is written as follows.

And in Fig. 7.7, under "scanning the barcode:"

Language used is very simple, and it adheres to British English spelling style. Example are words such as "colour" and "centres." Sentences are very long. The examples above attest to this fact.

More so, all four textbooks state that when writing instructions, it is necessary to state warnings clearly. Even though warnings are mentioned, they are not emphasized. Sometimes it is hard to tell if they are warnings, and special sections not created for warnings. They are interlaced with the prose of the text. Much importance is not placed on warning signs, and they are not clear. For example, the warning that cautions "users of the barcode scanner…" in Fig. 7.8 is too wordy and it is written in black ink; and the section does not present any sense of urgency.

Again, these sections "message requiring cleaning" (p. 8); maintenance messages of the BVD" (p. 6); "Low battery power" (p. 6); and "sleep mode of the BVD" (p. 4) captured in Figs. 7.3, 7.4, and 7.6 should be captured as warnings and cautions but they are not. Every sentence point to the BVD. What the manual is interested in is the biometric device and nothing else.

The user manual which accompanied the biometric device, I would argue, is a combination of what Seigel (2013) terms "system constitutive" and "system maintaining" documentation. By "system constitutive," Seigel refers to the tendency of a manual to "…argue for the future establishment of a technological system, or adoption of a technology, as a solution to an ideological, political, social problem" (p. 41). The user manual under discussion is system constitutive in the sense that it persuaded Ghanaians to have confidence in the device. Ghanaians were persuaded to believe it was necessary to adopt and use the device. If Ghanaians wanted a credible and accurate election results, then it was

HOW TO SWITCH ON/OFF THE BVD.

By pressing the *Power Button* once the following will be observed:
- The LEDs (amber colour) will switch on for a short period
- Start-up of device internals, with no visible action.
- Wait for 5 seconds
- Ghana Electoral Commission logo will be displayed
- While logo remains visible the device will continue to activate
- There will be a transition to "Ready" screen.
- After this the device is ready for use

To switch off the device do the following:
- Press the *Power Button* for about 4 seconds and release it
- The Red LEDs will blink
- *The BVD will switch off.*

Power Button

Sleep Mode of the BVD

In order to save batteries, the BVD automatically goes into *Sleep Mode* if not used for a given time. The BVD will move to *Sleep Mode* as follows:
- If there is no activity for 15 seconds the display will show a countdown timer and notify the operator that the device will go to 'sleep' in x seconds
- If the operator does not push any button while countdown timer is visible, the green LEDs will blink and BVD will enter sleep mode.

The BVD can enter *Sleep Mode* from the following situations:
- From the Ready Screen
- From the Scan Voter Register Barcode screen
- From the Verified and Rejected screen

Fig. 7.6 Instructions on how to switch the device on/off

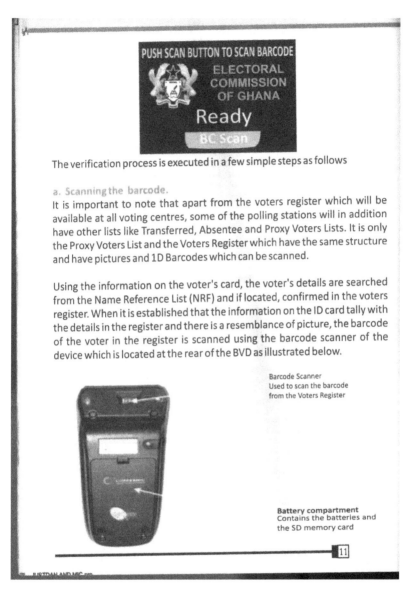

The verification process is executed in a few simple steps as follows

a. Scanning the barcode.

It is important to note that apart from the voters register which will be available at all voting centres, some of the polling stations will in addition have other lists like Transferred, Absentee and Proxy Voters Lists. It is only the Proxy Voters List and the Voters Register which have the same structure and have pictures and 1D Barcodes which can be scanned.

Using the information on the voter's card, the voter's details are searched from the Name Reference List (NRF) and if located, confirmed in the voters register. When it is established that the information on the ID card tally with the details in the register and there is a resemblance of picture, the barcode of the voter in the register is scanned using the barcode scanner of the device which is located at the rear of the BVD as illustrated below.

Fig. 7.7 Instructions on how to scan the barcodes on register with biometric barcode

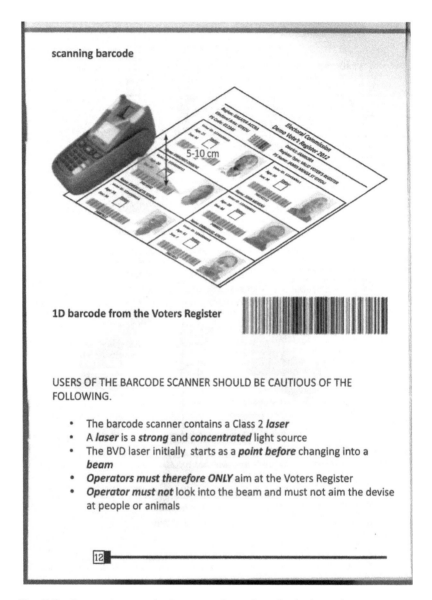

Fig. 7.8 Instructions cautioning on staring at laser in the barcode

necessary to participate and engage in this technological enterprise. Also, the user manual could be categorized as a 'system maintaining manual' document because it works to condition users to behave in a certain way. That is, it keeps "users engaged with a particular technology or system rather than to critique or change that system" (Seigel, 2013, p. 71). These can be identified through the rhetoric that surrounded the biometric machine (more on this topic in Chapter 5). By studying the manual, we get to understand why the country adopted the biometric technology, or better still, the context of use. It is also possible to conjecture through the analysis that users could not follow the instructions because the user manual at some point did not clearly state procedures. If an image comes without a text, as exemplified in Figs. 7.4 and 7.5, it can lead to misinterpretation or miscommunication. The user also spends a lot of time trying to figure out what the image means and how relevant it is to the process. During my interview with verification officers, I asked them to tell me what images in Figs. 7.4 and 7.5 stood for and it took them a maximum of 3–4 minutes to tell me those images stood for "handle with care" and "laser beam," respectively. Some could not tell me what the images were. Such a manual is likely to be ineffective and irrelevant to the user. The user might want to take charge of the process by abandoning the manual. This action on the part of the user could lead to the breakdown of the technology. As I indicated in Chapter 3, one of the Electoral Officers I spoke with indicated that the biometric devices broke down because they identified that some Data Clerks did not follow the instructions that came with the biometric technology. Verification officers would rather learn more about the system by doing than by referring to the manual. In the process, most of the machines broke down. The consequence of the breakdown was the extension of elections to the next day. This phenomenon somewhat confirms Johnson's (1998) argument that in a lot of cases it almost impossible for users to read instructions because they "are actively engaged with a technology that is taking all of their concentration" (p. 44). The solution to this problem, he identifies, is to "have instructional texts that are visually sensitive to the users' needs as they negotiate the activity simultaneously learning the technology while 'reading' the instructional text" (p. 44).

The Technological Gap

James Paradis is right in saying that documentation mediates the gap between expert and users in a technological system. Acknowledging that there is a gap indicates that in some cases users will struggle with technology use and that they have to make some adjustments before they can use adopted technologies. A typical example of such adjustments is captured in the epigraph that opens this chapter. When Ghana adopted the biometric technology, the EC realized that the user manual which accompanied the technology did not capture local needs so they had to redesign the document. For me, Ghana's example indicates that such concepts as localization and user-centered design still remain at the theoretical level. Ghana uses a technology developed in a European country. There was no contact between the producer of the technology and the real users of the technology. More disturbing is the idea that though the biometric technology was developed in Europe, the user manual that accompanied it was designed in Ghana by officials from the Electoral Commission and STL. These designers of the document may have not had any idea of how the biometric was even developed. At the mention of this, the first idea that comes up is Huatong Sun's (2012) revelation that there is almost always a disconnection between users and designers. The core of Sun's scholarly works contend that localization should lead to an understanding of use activities in a context. She makes us understand that a gap exists between the product designer and the user of the product. This gap is as a result of the existence of two levels of localization: localization at the developer's site and localization at the user's site. The tension between localization on the two levels should lead to the development of "an effective way to address cultural issues in IT localization and design well-developed products to support complex activities in a concrete context" (Sun, 2004, p. 2). In "The Triumph of Users," for instance, Sun admonishes us to make conscious efforts to study user localization, that is, efforts made by users to "integrate…technology their daily lives and enhance their lifestyles" (Sun, 2006, p. 474). In this direction, Ghana's case becomes a useful way of thinking about localization because there is an incorporation of efforts that users make to localize the biometric. This is seen in how officials of the Commission claimed agency by designing the user manual that accompanied the technology. Ghana's case can best be exemplified graphically in Fig. 7.9.

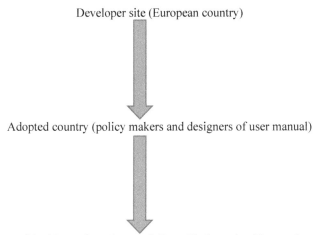

Fig. 7.9 Linear model of the user manual design process

In the diagram above, we see a confirmation of Sun's argument that there is almost always a gap between developer sites and user sites. These two sites are almost never in interaction. The best way to resolve this is to link "use back to design." She contends that users understand their use situations more than anyone else; hence, localization should take into cognizance "users at the use phase or expanding the design process to users' sites" (p. 477). This way, we can understand "concrete use activities in local contexts while understanding cultural and structuring factors" (ibid.). I agree. Similarly, Agboka (2013) advocates a "participatory localization" in the design of documents and technologies in international or cross-cultural contexts. By "participatory localization," the author refers to the extent to which users will be involved in design phases. The user who participates in localization processes is not an isolated person, but he/she is involved as a member of a community (Agboka, 2013, p. 42). He identifies that though technical communication scholarship is replete with ideas and notions of localization, little is done to account for user needs in *unenfranchised and disefranchised* cultural sites. Localization "often starts from the developer's site and trickles down to the users' site." As a result, "users are not cast as agents who initiate and implement change themselves" (p. 30).

We also have a glimpse of the role that power plays in the localization process. As the diagram above indicates, the user manual which intends to localize the biometric technology was designed by officials of the Electoral Commission. Polling officers who will use the document to help voters negotiate the voting process are never part of it. Polling officers are left out mainly because most of them are not permanent workers of the Commission. They are hired during general elections. Their jobs end as soon as elections are over. Thus the polling officer will almost always be sidelined in such design processes because he/she has a fluid identity. The interesting thing is that even though the Commission organized several training sessions for data clerks and other polling officials, most of them could not properly handle the biometric. Hence, the report that "the Commission had ... challenges such as the failure of some Data Entry Clerks to follow simple operational instructions regarding the usage of the biometric machines...."[1] Ghana's case is one of the many situations where users will not come into contact with developers of the technologies they adopt and use. The gap between producer sites and consumer sites keeps widening especially when user site is an unenfranchised cultural site that depends on donor agencies for economic support.

EXPERT DESIGNER TO THE RESCUE OF NON-EXPERT USER

A common trope in localization literature is that as global economies merge, it has become necessary for producers of technologies and artifacts to localize their products *to meet the needs of users* (Hunsinger, 2006; Salvo, 2001; Thayer & Kolko, 2004). For instance, Thayer and Kolko (2004) identify three levels of localization that takes place in the game industry: basic localization, complex localization, and blending. The first involves the translation of texts to meet user needs. However, the graphical user interface and game icons are left the same; in the second, graphical user interface, icons, and texts are translated to meet user needs. At this level, localization "focuses on the user interface and the functionality of the underlying code in the presence of the target language" (Thayer & Kolko, 2004, p. 18); and in blending, producers resort to two different options, they either "retain the narrative and simply translate the user interface details as well as possible, or rewrite the narrative to be more familiar to the players in another country" (ibid., p. 21).

Kirk St. Amant, Huatong Sun, and other localization experts identify that localization happens because the culture of use at the designer level is different from the culture of use at the user's' level; thus, such products designed may not be adequate to solve problems at the users' context. A situation Sun (2012) theorizes to mean two forms of localization processes: localization at the designer level and localization at the user level. This gap is the source of breakdowns in technological systems, she indicates. Therefore "items designed for one culture need to be revised to meet the preferences and expectations of another culture" (St. Amant, 2017, p. 114). In fact for a document or technology to be successful in other contexts, a redesign is quintessential. As has been the practice, organizations may "revise an existing product for international audiences or they may develop a new version of the product for audiences" (St. Amant, 2017, p. 114). These practices have given rise to terms such as *transcreation* and *glocalization*. On a more theoretical level, Kirk St. Amant (2017) suggests to technical communication scholars and information designers to combine script theory and prototype theory to help them understand "the expectations of other cultural audiences," (p. 114) so that they can design for effective use in other contexts.

Non-expert Users Rescue Themselves

I would want to comment, however, that with the exception of few scholars such as Huatong Sun and Godwin Agboka who see users as active agents, the main agent to initiate a redesign or revision process championed by localization experts is the designers of the technology. Designers are constantly called upon to revise their products to meet the expectation of consumers in other parts of the world. The developer of a technology is cast as the "heroic figure [who] is enlightened, principled, and capable" (Spinuzzi, 2003) and is able to combine theories to salvage breakdowns in use situations. The impression is that designers almost always "rescue" workers who are tyrannized by autocratic, despotic and unusable technologies; but in a lot of cases, users are able to rescue themselves; they are able to subvert autocratic documents that do not work for them. For this reason, subversive localization identifies a different worker or user of technology. This user is someone who is able to rescue himself/herself. This user does not wait to be rescued by the heroic designer. The quotation that opens this chapter is a tacit example of how users subvert technologies or documents. As was revealed by

Agorkoli (my interviewee), when the biometric technology arrived, the Electoral Commission realized that the user manual that accompanied the technology did not say anything about how to use the biometric in the electoral process. I am confident this problem occurred because the EC did not have any contact with the designers of the technology (as I indicated in Chapter 1, the EC officials I interviewed were not able to name the organization which designed the biometric technology the country used); neither was the biometric technology adopted designed to be used for election purposes. So with this problem at hand, who was the EC going to talk to? Who was coming into redesign the user manual? The EC thus had to rescue itself by designing or redesign a new user manual which will enhance and properly discuss how biometric technology can be used. Therefore, through this creative prowess, the EC "subverts the information system [which accompanied the biometric machine], inventing [its] own ways to turn" the biometric to their needs (Spinuzzi, 2003, p. 2).

Scholars such as Spinuzzi (2003), Sun (2012), and Johnson (1998) have realized that users create their own "practices, tools and texts constantly, sometimes in cooperation with the existing information systems, sometimes in competition with them" (Spinuzzi, 2003, p. 2). Take for instance Barbara in Spinuzzi's *Genre Tracing* who on realizing that a DOS-based database she was supposed to use to locate and analyze traffic accidents in an area in the Midwestern region in the US proved to be a cumbersome process to follow and ends up avoiding the software and instead using a Post-It note she had developed some months earlier; or Sophie, Lili, Brian, Mei, or Emma, characters in Sun's *Cross-Cultural Technology Design* who adopted the mobile technology and localized it to fit their local needs. In all of these examples, documents imposed on users or technologies adopted did not capture "local contingency" and so the users created a "work around" (Spinuzzi, 2003, p. 3) or they found a way out. "Finding a way out" is core part of subversive localization. It is the reason I support calls that emphasize the need to pay attention to local use situations and how users initiate innovative ways to salvage themselves. If users find a way, then it means they have specific kind of knowledge which must be observed and respected.

If users have the power to subvert and initiate creative ways to save themselves, then why are they portrayed as idiots or victims waiting to be rescued? Johnson helps us to think about this damning perception of users by indicating that this is because we have not properly interrogated "use" and the knowledge making an aspect of it. We need to "understand the knowledge that users have of technological artifacts and systems;" "we need to pay serious attention to the nature of user knowledge" (Johnson, 1998, p. 45). He walks the talk by placing users' way of knowing into three categories: user as a practitioner, user as a producer, and user as an active citizen. In the first, the user is perceived as a mere tool user. The user in this regard, only puts already designed technology to use; in the second, users do not merely use but they contribute to the design and maintenance of the technology; and in the third, user serves as an active participant in a democratic society (Johnson, 1998, pp. 46–60). Although all three categories help to capture the traits of a user who embodies knowledge, I would like to focus on the user as a practitioner. In Johnson's discussion, he mentions that the user as practitioner is not a mere user of technology (as has been conceived throughout technological history), but the practice aspect of use helps to flesh out the classical Greek term *techne*—the art of making (p. 51). A brief detour will be appropriate.

User Rescues Through Practical Knowledge

According to Johnson, *techné* has received a lot of attention from classical scholars such as Plato and Aristotle. Chief among the arguments surrounding this term is whether or not there is a "knowing" with *techné* or simply put, "is *techné* epistemic?" (p. 51). Aristotle also makes a distinction between craft-knowledge (*techné*) and scientific knowledge (episteme). In book VI of Nichomachean ethics, Aristotle makes us understand that scientific knowledge is concerned with what is necessary. Scientific knowledge is the search for eternal knowledge. Techné, according to Aristotle, is concerned with production. It is concerned with things that come to be (Aristotle, 2014). This is unusual to read in classical Greek scholarship the association of *techné* with carpentry, navigation, weaving, or fishing.

Perhaps Carl Mitcham can help us to identify that techne has a knowing to it. Techne is commonly used in Greek to refer to "art," "craft," or "skill," and it is used literally to refer to "cleverness and cunning in getting, making, or doing as well as to specific trades, crafts and skills of every kind" (p. 117). He identified the use of techne in both Platonian and Aristotelian senses. In Plato's conception, techne is linked up with episteme, that is, there is a relationship between art and systemic or scientific knowledge. Every techne "is involved with logoi (word, speech, reason). Plato identifies two senses of the word "techne." First, techne is used to refer to work that involves physical activities and it requires the use of language. Techne, as used in the second sense, refers to activities that involve less physical activity, like say arithmetic. What binds both senses is the idea that "techne refers to all human activities that can be talked or reasoned about" (p. 118). This is extended in the classification of knowledge by Plato: knowledge which stems from *education* and upbringing and knowledge involved with making or producing. He identifies that Aristotle also uses *techne* to refer "to a special knowledge of the world that informs human activity accordingly" (p. 120). Mitcham identifies that techne as used by Aristotle forms a continuum that moves from sense impressions and memories through experience to systematic knowledge, episteme. That is a distinction between art (techne) as a form of bringing forth, or coming to be and episteme (science), the realm of being. Techne is not just an ordinary activity, it is "a capacity for action, founded in a specialized kind of knowledge…a habit with true logos concerned with making" (p. 120). Thus techne, is, in one sense, epistemic because it involves the use of consciousness about the world; it can be taught or communicated. In another sense, it is not episteme because it deals with what is contingent, and it depends on things that change rather than the unchanging. Although techne is a devalued form of knowing associated with users, it provides a "rule of thumb for solving a problem, even if the solution is only a temporary fix" (Johnson, 1998, p. 52). He believes that through techne, users are able to develop varied strategies they fall on to solve their own problems. They do not have to wait for designers or experts to come to save them from the abyss of distraction. In this regard, users as practitioners embody "cunning intelligence" (p. 46), an idea linked to the mythic figure metis.

Johnson is not the only person who establishes a relationship between techne and metis. Atwill (1998), for example, state metis and Kairos as some of the terms which "define[…] techne as a model of knowledge with a distinctive form of intelligence and sense of time" (p. 48); and Dolmage establishes a close-knit relation between metis, techne, Kairos, and tuche (luck). They work together in all circumstances. In this regard, to resurrect user knowledge and to portray users as active agents who create their solutions, we must draw cunning connections between the user and metis. What traits of metis do users who subvert documents possess? And why will an attention to metis help to recover the lost voices of users?

Like users, metis, though powerful, has been relegated to a subservient position in Greek myth. For instance, the all-powerful goddess is swallowed by Zeus and Detienne and Vernant state that after all the traits that metis possesses, "its sleights of hand, its resourceful ploys and its stratagems, are usually thrust into the shadows, erased from the realm of true knowledge and relegated...to the level of mere routine, chancey inspiration...or even charlanterie..." (p. 5). In fact, Dolmage (2009) tells us that metis is relegated for a reason; "we have erected a rhetorical tradition that valorizes the split between the mental and the physical...the body and the soul" (pp. 2–3) or the split between rhetoric and philosophy while Detienne and Vernant (1991) identifies a split between "being and becoming, between intelligibility and the sensible" (p. 5); and because metis is associated with bodily intelligence, becoming, and the body; domains that are seen as subpar to sensibility, being, and the soul, "metis has been subject to derogation" (Dolmage, 2009, p. 5). In spite of this denigrated status, metis, the wife of Zeus, is described as a manifestation of a complex mental attitude that exhibits flair, wisdom, forethought, subtlety of mind, deception, resourcefulness, vigilance, opportunism, and various skills and experience acquired over the years (Brady, 2003; Detienne & Vernant, 1991; Dolmage, 2009). Dolmage (2009) states that metis is "cunning, adaptive, embodied intelligence...In Greek, metis means wisdom, wise counsel—but it also means cunning and connotes trickery" (p. 5), or "the capacity to act in a kairotic world" (Dolmage, 2006, p. 121).

In Classical Greek mythology, the cunningness of Metis was used in defeating the Titans by the Olympians; Her powerfulness is celebrated by Detienne and Vernant when they assert that "without the help of the goddess [Metis], without the assistance of the weapons of cunning, she controls through her magic knowledge, supreme power could neither be won nor exercised nor maintained" (p. 58); Metis is relational, never works alone but with the help of three powerful forces: tuche (luck), kairos (opportunity to succeed), and techne. Metis is timely; it is flexible and very practical (Dolmage, 2006, p. 121). Thus the person with "metis perceives the world of tuche, harnesses kairos, and has the ingenuity required to think of cutting and building the tiller" (Dolmage, 2006). Metis is usually associated with the weak in the society; Hephaestus (dolmage) and Antilochus (Detienne and Vernant) are but a few examples. In most instances, those who are disadvantaged devise strategies in order to turn tables around. Hephaestus, for example, who was banished from Olympos, for reasons not tacitly known (see Dolmage, 2006 for more explanations), returns to Olympos years later as the "god of fire, a gifted craftsman" (Dolmage, 2006, p. 127) who builds "...each of the gods a house...a golden breastplate for Heracles, armor for Achilles...a home and a house for Zeus" and makes the all-powerful Pandora (Dolmage, 2006, p. 129). As Dolmage indicates, Hephaestus is able to do all of these because he enacts metistic traits. Also, by enacting metistic tendencies, Antilochus, who was seen as the least favorite to win a horse race, managed to outstrip his contenders who had horses that could run faster than Antilochus'. Detienne and Vernant indicate this success is possible because of metis, which they link to "informed prudence" (p. 26). In fact, Detienne and Vernant never shied away from firing a connection between metis and prudence. In Chapter 1 of *Cunning Intelligence*, the authors defined metis as "a particular type of intelligence, an informed prudence..." (p. 11). This means that metis and those who exhibit metistic tendencies embody some kind of practical wisdom/knowledge, "a type of intelligence and of thought, a way of knowing..." (p. 3). This kind of prudence or practical wisdom helps them to find a way.

A redesign of the user manual by users who were not part of the design of the biometric technology, I will argue, is an exercise of cunning intelligence. It is a testament to the fact that the person who is endowed with metistic tactics is ready to judge the moment to initiate solutions; that user acts faster to find a way or "work around" as Spinuzzi indicates. This idea that users embody practical knowledge supports my argument that in redesigning the user manual, the EC and officials exhibit this kind of cunning intelligence. For how else could they have designed a user manual for a technology they never developed? In essence, they managed to judge the moment to act. The numerous examples stated in this chapter indicate the extent to which users embody cunning intelligence that enables them to subvert and redesign documents or technologies to meet their needs. If we acknowledge that techne embodies episteme then we see how relevant it is to study the various ways users subvert technology. I argue that if we need to really demonstrate and defend the notion that users are creative and knowledgeable, then we need to study the various ways they subvert documents imposed on them. By subverting, they demonstrate cunning intelligence; they are able to judge the moment and act to save themselves. Apart from the fact that the user manual speaks to the practical knowledge of users, the manual also helps to capture local logics and in the next section, I indicate how this is possible.

Manual as an Articulation of Local Logics in a Context

An attempt to understand the manual leads to the recognition that rhetorical and cultural patterns of a context influence, in a way, how documents are designed. This confirms earlier studies that make similar claims. Scholars, such as Thatcher (2006, 2011), Fukuoka et al. (1999), Agboka (2012), G. Y. Agboka (2013), Boiarsky (1995), stress that documents reflect the rhetorical and cultural patterns of societies. Thus, how does this manual articulate a cultural and rhetorical value of a non-Western society? To respond to this, I turn attention to Thatcher's (2011) revelation that one of the ways to assess manuals for implicit cultural and rhetorical values is to describe the instructional manual based on "eight common human thresholds" (p. 45), namely levels of independence or interdependence among human groups (I/other); how cultures develop and enforce rules (rules orientation); definition of boundaries among what is public or private (private/public); variety of sources that guide behavior (sources of virtue and guidance); status and accomplishments; context in communication; how cultures deal with time (Time); and how cultures deal with inequality (Power distance). These eight overarching thresholds reflect or reveal how cultures identify the self, ideologies or social values, epistemologies or thinking patterns and subjectivity concepts I stated in Chapter 3. More so, the thresholds identify how rhetoric acts upon culture in a context. Thus the thresholds enable an understanding of the dialogic and articulated relationship between culture and rhetoric—a revelation that helps understand how rhetorical–cultural methodology can be operationalized to conduct international technical communication research. Although all eight thresholds are relevant, I focus on how the user manual articulates four of these units: I/Other, norms or rules, status, context, and power distance.

Common Human Thresholds and Their Relevance to Document Analysis

The I/Other threshold establishes the importance of human interaction. Essentially, it hinges on the idea that humans are social beings and thus interact with individuals or groups. This threshold, according to Thatcher (2011), "assesses the levels of dependence or independence among people" (p. 46). Basically, this threshold categorizes societies as either Individualistic or Collectivistic. In an Individualistic society, individuals see themselves as independent and as a result, see the world through "individual identities and efforts." This sense of individualism is captured in their communication patterns, mostly through the use of "I," "a strong bifurcation between personal and objective communication strategies; a dumbed-down readership level so as to un-complicate the interpersonal dependence of communicators; and an emphasis of personal achievement, self-creation, and reader-friendly document design patterns" (p. 47). In a Collectivist society, on the other hand, individuals view themselves as dependent on others, and thus, define themselves as part of a group. In these collective societies, "communication patterns emphasize interpersonal relationships, social hierarchy, social leveraging, group identities, close personal space, and writer-friendly patterns" (p. 47).

As a born and bred Ghanaian, I can say that Ghana is a Collectivist society. As Utley (2009) puts it, the social values of Ghana place importance on "social harmony and the well-being of others, not just of the individual" (p. 40). Growing up in Ghana, for instance, I have come to cherish the importance of the extended family and the community. Family in Ghana does not refer to mother, father, and child but a close-knit relationship between siblings, cousins, grandparents, and other relatives. It is common for a Ghanaian child to grow up in a house that consists not only of their siblings but other relatives. I, for instance, did not grow up with my parents. At an early age, I was taken up and nurtured by my uncle who has four children. My uncle's wife also had a sibling living with us, and constantly family members from both sides visited. Therefore, at an early age, the notion that family is about people and not just siblings was drummed into my head. But how is this Collectivist value reflected in the design of the user manual? As I stated in the preceding paragraph, the difference between collective and Universalist values can be elucidated through assumptions made about readers or writer's rhetorical strategies. Documents are either reader-centered or writer-centered. Reader-centered writing patterns are linked up with Universalist (Individualist) cultures where communication relies "on overt visual designs such as headings, hanging indents, concise introductions and conclusions, lists, and other formatting devices that signal the content of instructions through visual design" (p. 288). Whereas Collectivist cultures rely on "the complexity of the author's thoughts with little effort to revise the writing or design the document to accommodate different levels of audience knowledge and attitudes"

(p. 289). Writer-friendly rhetorical patterns, Thatcher continues, draw "on the same tacit knowledge, non-verbal and contextual cues, and other rhetorical strategies and processes known by the in-group" (ibid.).

The user manual, I would say, is a combination of both reader-centered and writer-friendly approaches. At some point, the manual relies on overtly visual designs and visual cues, communication patterns of Universalist cultures. There is the preponderant use of headings and lists, and readers are babysat throughout the document. Almost every page of the document has some form of a visual cue. Nineteen out of the 23 body text pages have a visual cue of some sort. For example, visual cues that indicate the oblique view of the device (Figs. 7.1 and 7.10).

Or cues that indicate how an individual should place their fingers on the device. An example is Fig. 7.11.

In essence, there is "sufficient information to accomplish all that is necessary..." (p. 289). On another level, there are traces of writer-friendly communication approaches. For instance, the manual makes explicit statements about the audience of the document: operational staff, such as key trainers, polling staff, and stakeholders who will handle the biometric machine on Election Day. There are also instances where the manual adopts a conversational approach, indicating an oral form of communication. An example is Fig. 7.7. And in some sections, images do not come with texts that explain what those texts are (refer to Figs. 7.4 and 7.5).

The combination of Collectivist and Individualistic rhetorical forms gives an indication that there are instances where some cultures, such as Ghana, will adopt both Universalist and Collectivist values in their communication patterns. This, in a way, identifies the dynamic nature of the Ghanaian society.

"All cultures establish norms or rules," states Thatcher's second human threshold. But Thatcher is quick to point out that the manner in which these rules are laid out vary from culture to culture. Mostly, these rules or norms adopt a universal or a particular approach (p. 48). In a Universalist culture, laws and regulations adopt a more "level playing field" approach. Regardless of the situation, laws and norms are enacted based on what is good and right for the society and not the social status of individuals. It operates under the dictum: All are equal before the law. Communication patterns include "strategies of fairness, justice, equality, and parallelism, and strong use of templates or branding" (p. 48). In a particularist society, however, rules depend on "relations and context."

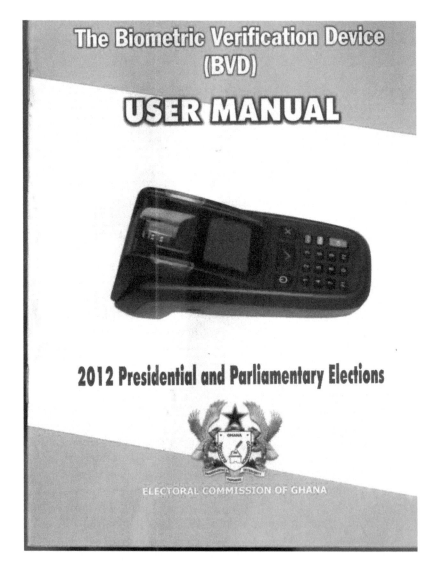

Fig. 7.10 The oblique picture of the biometric device used during the elections

(c). **Showing Verification Results**
- The results of the scanning are shown below with an **Ok** (√) green sign or **Rejected**(x) red sign.

End of day procedure

At the end of the day, the *number of Successful Verifications* must be handed over to the Presiding Officer after which the device will be switched off.
- The number of *Successful Verifications* can be retrieved by pushing the F2 button.

BVD Display	Polling Officer action
PUSH SCAN BUTTON TO SCAN BARCODE / ELECTORAL COMMISSION OF GHANA / Ready	• Push the F2 button on the keyboard
Verification Summary / Number of Successful Verification: 18 / Return	• Record the number of Successful Verifications and hand over to the Presiding Officer • Push Return to move the Ready screen

TYPES OF VOTERS LISTS AVAILABLE AT POLLING STATION
Each polling station will have the following
- The Main Voters Register.
- The Names Reference List

Fig. 7.11 Instructions showing possible outcomes of verification

Common among particularist communication patterns are "emphasis of context and particular relationships; attention to exceptional circumstances; and consideration of social prestige and power relations" (p. 48). Related to norms and rules is the fifth threshold: status. This threshold pushes us to understand how status is developed and communicated in a society. Here, scholars identify two approaches: ascription and achievement. In ascriptive-oriented cultures, status is derived from "social group, race, gender, age, ethnicity, and language" (p. 55). The communication pattern of ascriptive societies is "operationalized through overt signs of ascribed status, power through people, and importance of context and history to the university" (ibid.). In achievement-oriented cultures, however, status is not derived from where you come from, your family background, language, etc. Instead, an individual is okay as long as they can achieve their goals. Achievement-oriented cultures tend to emphasize mostly on what they have achieved when they are communicating. Achievement-oriented communication "emphasize action, change, statistics, and other observable and verifiable demonstrations of accomplishment" (p. 55).

It is not by mere coincidence that after the "foreword" section is written, the Chair signs his name as Dr. Kwadwo Afari-Gyan. Most Ghanaians will attest to the fact that titles are cherished in Ghanaian societies. It distinguishes and creates a hierarchy among individual members in the society, and it is "not unexpected for a society in which hierarchical relationships prevail" (Dong, 2007, p. 226). One of the cultural shocks I experienced when I arrived at Michigan Tech was the rate at which professors will ask students to refer to them by their first names. It is rare to see this attitude in Ghana. Friends have narrated stories where they have been asked to leave a lecturer's office because they did not address them by their proper title: "Doctor." The fact that the foreword was written by the Chair speaks to the level of seriousness that needs to be attached to the user manual and the rules that have been laid out.

Issues of norms and status lead to a discussion of another threshold: "power distance." This threshold pays attention to the various ways inequality is handled or revealed through communication. He identifies high power and low power distance approaches. High power societies use communication to reinforce and exemplify hierarchy in society. There is, as Hofstede identifies, an "unequal, centralized distribution of power, acceptance of authority, high levels of overseeing, and top down communication styles" (Thatcher, 2011,

p. 62). Communication patterns are usually top-down, strict, adhere to formalities and titles that indicate one's hierarchy in the society. In contrast, communication is less formal in a low power distance culture and it tends to ease off inequality.

I have discussed the preponderant use of visual cues and list in the user manual. Since I did not talk to users or observe them using the manual, I can only conjecture that writers assumed that by following the visuals and laid down steps, polling staff will be able to use the device to achieve set target. As indicated above, goal achievement is a trait in a Universalist approach to communication. Thus, it is believed that regardless of the educational or economic background of the polling staff, they will be able to use the manual effectively. The goal of the commission is to enable individuals to cast their votes on Election Day. Thus, every section of the manual gave information that is needed for the job to be done. As we can infer from most of the images above, the various processes and stages are broken down for the user to follow. For example, Figs. 7.12 and 7.13 lay out the "general flow in polling station." These pages lay down what polling agents are to expect in regard to the flow of voters in order to help them to anticipate what work schedule will look like. They have to know what will happen and what they need at each stage of the process.

I end this section with a caution: The fact that rules and norms are laid out for users to follow does not mean that the Ghanaian society follows a Universalist approach at every instance. In some instances, society adopts a particularist approach.

The final threshold to discuss is the context in which communication occurs. Thatcher distinguishes between low and high context of communication. Low-context communication places emphasis on explicitly written protocols of communication rather than surrounding context. As Thatcher (2011) identifies, in low-context communication societies, attention to detail is high and communicates "all of the verbal signs and cues to carefully guide the reader through the message" (p. 56). High-context communicators, on the other hand, depend on social cues to ascertain the meanings embedded in communication. "Good writing focuses on information that elucidates the influence of the social context on meaning..." (p. 57). High-context communicators hardly rely on fixed guidelines. They tend to cherish non-verbal, indirect approaches to communication. Instances cited above demonstrate, again, that communication patterns of the Ghanaian society shift between high

The undermentioned lists may not be available at certain centers
- Proxy voters List.
- Transferred Voters List.
- Duplicates List
- Exceptions List
  As indicated earlier, only the Voters Register and the Proxy List have barcodes for scanning.

THE VOTERS REGISTER

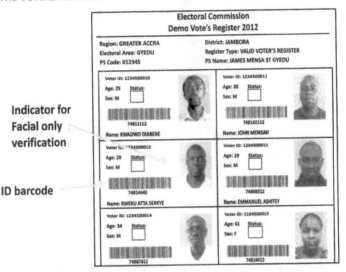

General Flow in Polling Station
- **Voter** enters Polling Station and presents Voters ID Card to the polling assistant in charge of the Name Reference List (which is arranged alphabetically to make it easier for the voter's details to be located). The Polling Assistant searches for the name and if successful, communicates to the Verification Officer the page number and column where the voter's details can be found in the

Fig. 7.12 The voters register in view

voters register. Polling Assistant then marks the voter's name on the list as having been found.

- ***The Verification Officer*** finds the entry in ***the*** Voters Register using the information provided by the Names Reference List, and the details on the voter ID card.

- The verification officer makes a determination of identity (picture on card, picture in register) and if the results are positive the barcode of voter is scanned after which the applicant's finger(s) is scanned.

- If the Biometric fingerprint scanning is ***successful***, the voter is inked and issued with a ballot paper to vote. Verification officer then marks the voter's barcode in the register to indicate voter has been verified.

- If the Biometric Verification is ***not successful***, the presiding officer will be alerted and he/she will take a final decision on the voter.

- In some instances the device will verify the voter by face only. This happens when none of the voter's fingers was captured during registration.

NB: It is only a barcode from the Voter's Register that should be scanned. Voter Identification Cards should not be scanned

Fig. 7.13 Verification procedures for verification officers

context and low context. For instance, every rule or norm necessary for successful use of the biometric and a successful election is coded in the user manual. Apart from the steps and laid down procedure, polling staff are charged to disallow "a voter whose fingerprints are not recognized by the device." Meanwhile, "voters whose fingerprints are recognized by the device will be issued with ballot papers" to cast their votes (manual, p. 1). However, there are instances where users would have to rely on the situation or external environment to make a decision. For instance, when a voter's fingerprint is rejected, do you send the person home instantly considering the fact that the person has been in a queue for close to 10 hours? It is important to stress that even though official voting starts at 7:00 a.m. GMT, some voters wake up as early as 1:00 a.m. to queue (some even start queuing a day before election). Will a polling officer have to ask someone who queued from 3:00 a.m. to 2:00 p.m. to go home just because the biometric device rejected the individual? When such situations occur, the "presiding officer will take a final decision on voter eligibility" (manual, p. 20) by asking voters around to ascertain the voters state. At this point, the decision is made outside of the coded, written guideline. The idea here may be that by using common sense or by depending on the context to make a decision, polling officers will be able to solve any problem they encounter during the electoral process.

These thresholds make us understand that "cultural [and rhetorical] differences are indeed a relevant factor to consider when addressing international audiences in any form of international communication," (Dong, 2007, p. 222). The thresholds could also be a way to understand why some of the sections deviated from the guidelines to instruction writing we teach in US technical communication classrooms. They provide a window to appreciate the essence of understanding how diverse cultural and rhetorical patterns shape document design. In this regard, when a document from another culture deviates from what we are familiar with, we are not quick to identify it as worthless. Such documents from other cultures help us to reflect on our encounter with technical documents from cultures different from ours. In the next chapter, I spend enough time discussing the social justice implications of the biometric technology adopted and used in Ghana.

Note

1. http://www.ec.gov.gh/news.php?news=116.

References

Agboka, G. Y. (2012). Liberating intercultural technical communication from "large culture" ideologies: Constructing culture discursively. *Journal of Technical Writing and Communication, 42*(2), 159–181.

Agboka, G. Y. (2013). Participatory localization: A social justice approach to navigating unenfranchised/disenfranchised cultural sites. *Technical Communication Quarterly, 22*(1), 28–49. https://doi.org/10.1080/10572252.2013.730966.

Aristotle. (2014). On "Techne" and "Episteme". In R. C. Scharff & V. Dusek (Eds.), *Philosophy of technology: The technological condition: An anthology* (2nd ed., pp. 19–22). West Sussex: Wiley Blackwell and Sons.

Atwill, J. M. (1998). *Rhetoric reclaimed: Aristotle and the liberal arts tradition*. Ithaca: Cornell University Press.

Baneseh, M. A. (2015). *Pink sheet: The story of Ghana's presidential election as told in the Daily Graphic*. Accra: G-Pak Limited.

Bazerman, C. (1998). The production of technology and the production of human meaning. *Journal of Business and Technical Communication, 12*, 381–387.

The Biometric Verification Device User Manual. (2012). In T. E. C. o. Ghana (Ed.). Accra.

Boiarsky, C. (1995). The relationship between cultural and rhetorical conventions: Engaging in international communication. *Technical Communication Quarterly, 4*(3), 245–259.

Brady, A. (2003). Interrupting gender as usual: Metis goes to work. *Women's Studies, 32*(2), 211–233.

Detienne, M., & Vernant, J. P. (1991). *Cunning intelligence in Greek culture and society*. Chicago: University of Chicago Press.

Dolmage, J. (2006). "Breathe upon us an even flame": Hephaestus, history, and the body of rhetoric. *Rhetoric Review, 25*(2), 119–140.

Dolmage, J. (2009). Metis, Metis, Mestiza, Medusa: Rhetorical bodies across rhetorical traditions. *Rhetoric Review*, 1–28.

Dong, Q. (2007). *Cross-cultural considerations in instructional documentation: Contrasting Chinese and US home heater manuals*. Paper presented at the Proceedings of the 25th Annual ACM International Conference on Design of Communication.

Fukuoka, W., Kojima, Y., & Spyridakis, J. H. (1999). Illustrations in user manuals: Preference and effectiveness with Japanese and American readers. *Technical Communication, 46*(2), 167–176.

Gurak, L. J., & Lannon, J. M. (2012). *Strategies for technical communication in the workplace*. New York, NY: Pearson Higher Ed.

Hogan, D. B. (2013). Theories that apply to technical communication. *Connexions: International Professional Communication Journal, 1*(1), 155–165.

Hunsinger, R. P. (2006). Culture and cultural identity in intercultural technical communication. *Technical Communication Quarterly, 15*(1), 31–48.
Johnson, R. (1998). *User centered technology: A rhetorical theory for computers and other mundane artifacts.* Albany: State University of New York.
Markel, M. (2013). *Practical strategies for technical communication.* Boston: Bedford/St. Martin.
McKeon, R. (2009). *The basic works of Aristotle.* New York: Random House LLC.
Oliu, W. E. (1994). *Writing that works: Effective communication in business.* Scarborough, ON: Nelson.
Paradis, J. (2004). Text and action: The operator's manual in context and in court. In J. S. Johnson-Eilola & S. A. Selber (Eds.), *Central works in technical communication* (pp. 365–380). New York: Oxford University Press.
Pfeiffer, W. S. (2003). *Technical writing: A practical approach.* Toronto: Pearson College Division.
Salvo, M. J. (2001). Ethics of engagement: User-centered design and rhetorical methodology. *Technical Communication Quarterly, 10*(3), 273–290.
Seigel, M. (2006). *Reproductive technologies: Pregnancy manuals as technical communication.* Ann Arbor: ProQuest.
Seigel, M. (2013). *The rhetoric of pregnancy.* Chicago: University of Chicago Press.
Spinuzzi, C. (2003). *Tracing genres through organizations: A sociocultural approach to information design* (Vol. 1). Cambridge, MA: MIT Press.
St. Amant, K. (2017). Of scripts and prototypes: A two-part approach to user experience design for international contexts. *Technical Communication, 64*(2), 113–125.
Sun, H. (2004). *Expanding the scope of localization: A cultural usability perspective on mobile text messaging use in American and Chinese contexts.* Troy, NY: Rensselaer Polytechnic Institute.
Sun, H. (2006). The triumph of users: Achieving cultural usability goals with user localization. *Technical Communication Quarterly, 15*(4), 457–481. https://doi.org/10.1207/s15427625tcq1504_3.
Sun, H. (2012). *Cross-cultural technology design: Creating culture-sensitive technology for local users.* New York: Oxford University Press.
Thatcher, B. (2006). Intercultural rhetoric, technology transfer, and writing in US–Mexico border maquilas. *Technical Communication Quarterly, 15*(3), 385–405.
Thatcher, B. (2011). *Intercultural rhetoric and professional communication: Technological advances and organizational behavior.* Hershey, PA: IGI Global Press.
Thayer, A., & Kolko, B. E. (2004). Localization of digital games: The process of blending for the global games market. *Technical Communication, 51*(4), 477–488.
Utley, I. (2009). *The essential guide to customs and culture: Ghana.* London: Kuperard.

CHAPTER 8

You Are Not Who You Say You Are: Discriminations Inherent in Biometric Design

My decolonial approach to localization will not be complete if I fail to discuss the social justice implications of the biometric technology adopted by Ghana. As I indicated in Chapter 3, one of the core goals of decolonial approaches and by extension, localization scholars, is to empower the marginalized in our societies. I reveal in this chapter that the use of biometric technology provides an avenue to examine the role technologies play in marginalizing users. Apart from the fact that the biometric technology broke down, interview analysis reveals another disturbing issue: The biometric technology could not read the fingerprints of some voters. Mr. Manu states:

> …what we saw from the use of the [biometric technology] is that we saw that people who had rough fingers as a result of the work that they do; especially farmers, mechanics or fitters or whatever, yes those who had rough fingers they were finding it very difficult to get verified. This is what we actually observed.

Abena Kwarteng also revealed that:

> During the training we were trying to see if maybe during the day of election everything will be okay so we had to test it for ourselves…What we realized was that some of them the finger print it couldn't pick the finger print….

Other sources, such as a report by the Commission on Human Rights and Administrative Justice (CHRAJ), revealed that "Most of those affected by the problems of verification were elderly women, some young women who had dyed their hands and fingers with a local herb called 'lenle,'" (*Preliminary Report on Monitoring the Right to Vote and Observing the 2012 Presidential and Parliamentary Elections in Ghana*, 2012, p. 6) and Kofi Ansah also revealed that "mostly the elderly, like farmers, fishermen and then these mechanics and all those things." It is interesting to note that a similar observation was made by Hobbis and Hobbis (2017) when they conducted an ethnographic research to examine biometric use in elections in the Solomon Island. In this study, they found out that the biometric could not read the fingers of individuals who were actively engaged in agriculture and or "non-industrial fishing" and those whose faces have gone through some form of medication "from aging through the application of cosmetics, plastic surgery or ...tattoos" (Hobbis & Hobbis, 2017, p. 118). These examples attest to the fact that biometric discriminates against a specific group of people, hence the need to discuss the social justice implications of biometric use in elections.

It is not the biometric which is rejected by voters; on the contrary, it is the people who wish to express their free will to vote that are rejected and discriminated against. This revelation is consistent with Magnet's (2011) claim that "human bodies are not biometrifiable" (p. 2) and that new identification technologies such as the biometric technology "suffer from 'demographic failures'" (p. 5). Magnet points out instances where the biometric has failed: biometric fingerprint scanners struggle to scan the hands of Asian women; Iris scanners exclude people in wheelchairs and those who are visually impaired; and worn-down or sticky fingerprints are rejected by fingerprint biometric scanners. The implication is that any type of body that falls outside the limit of the biometric lens cannot be accurately read or verified.

Murray (2007) extends Magnet argument by revealing that the biometric technology struggles to read all bodies in the same manner and that the biometric produces "illegible biometric bodies." This means that any individual body that is rejected by the technology becomes illegible because that body is not considered appropriate at the space it finds itself. So, for instance, if one is registered to gain financial assistance or welfare and at the point of biometric verification that individual is rejected, the person does not get access to welfare or financial assistance.

Or if an individual who is registered to vote is rejected by the biometric on Election Day, that individual does not get to vote, as was the case in Ghana in 2012 when biometric was introduced. Murray empathizes with those who are rejected by stating that they suffer rejection not because they are abnormal or that they have deviant or illegitimate bodies; it is the "technology that struggles to read them" (p. 351), an affirmation that the biometric technology is "gauged to the idealized bodies in a given culture, producing as 'abnormal' those who do not correspond to the idealized model….Biometric technology has been made, therefore, with a normative notion of 'body' in mind; a culturally constructed notion of embodied identity…" (p. 351). She goes on to argue that "attempting to translate a body into computer codes does not free it from the social codes that produce stereotypes" (p. 357); thus, those with "careers that require manual labor have abused hands" (p. 358) are rejected by biometric technology; those who have small hands (Japanese flight attendants have had issues at LAX), or plumbers, carpenters, bricklayers, and health care workers (p. 351) also suffer rejection.

Thus, as a human-centered or user-centered field, these pockets of injustice should propel us to question or interrogate why the biometric technology rejects a specific group of people, a phenomenon which prevents some individuals from casting a ballot to elect their representatives. As I indicated in Chapter 2, the Universal Declaration of Human Rights enshrines on individuals the right to vote or participate in electing representatives. In fact, if we will be true to Walton's (2016) call on technical communicators to support human dignity or human rights then we should turn attention to electoral systems to find out how technologies adopted to enhance electoral systems disenfranchise voters. In Chapter 1, I indicated that our field has taken an interest in discussing issues related to civic engagement (Bowdon, 2004; Cushman & Grabill, 2009; Dubinsky, 2004; Flower, 2008; Rude, 2004; Simmons & Grabill, 2007), but, conversation about elections, and especially, about the discriminatory tendencies inherent in election technologies, remains a less talked about area in TC. Therefore, in my quest to decolonize and expose biometric disenfranchisement, I ask: How does Ghana's case help to interrogate electoral disenfranchisement? Why does the biometric reject people with worn-out fingers?

Discriminations Inherent in (Technology) Design

Social justice and localization discussions are replete in the field of technical communication, but little is done to extend such conversations to colonized sites. In cases where such conversations have happened, such discussions take a top-down approach. Agboka (2013) argues that if we take a social justice approach to localization, technical communication, as a field, can address issues related to technology design, usability, information technology, and communication. I must say that research by Huatong Sun corroborates Godwin Agboka's call for the study of design concerns in non-Western contexts. For instance, in her research into the various ways, users of a mobile technology localized the device to meet their local needs, Sun (2012) identifies that technology design processes create "a disconnect between action and meaning" for local users (p. 8). That is, little is done to incorporate the meanings, behaviors, practices, and state of consciousness of users in cross-cultural contexts into technology design. She believes that technologies breakdown because of the disconnect that exists between users and designers. But, the question, who gets affected when a technology breaks down, is not addressed. The examples I have provided in regard to the rejection of voters by the biometric,are a tacit confirmation of the claims Godwin Agboka and Huatong Sun make. In Chapter 5, I examined the communication surrounding the biometric technology and found out that the biometric was cast in super-hero narratives; in Chapter 6, I examined a counter-narrative to the picture painted of the biometric in Chapter 5. I revealed how, despite the positive communication, the biometric broke down on several fronts. This chapter also adds to the biometric breakdown narrative by questioning why the biometric could only reject specific individuals. It is clear from the ongoing that localization of technology should be discussed in tandem with social justice concerns.

Some technical communication scholars who study the relationship between rhetoric, technical communication, and culture have identified the extent to which technology use can become a tool to upend race or discrimination (Haas, 2012; Jones, 2016a; Katz, 1992; Rose, 2016; Walton, 2016). Haas (2012), for example, indicates that "technologies are not transparent" objects but are "cultural artifacts imbued with histories and values that shape the ways in which people see themselves and others in relation to technology" (p. 288). This means that technologies are not neutral, and they reflect the culture in which they are designed.

I believe technology designers have come to know this as a fact, so they work hard to either internationalize, customize, or localize their technologies. Earlier, I indicated how technical communication scholars point out that these honest efforts by technology designers to fit technology into user contexts fail because the localization approaches do not account for user actions. Unfortunately, such same spirit is not invested to explore the extent to which the biometric disenfranchises voters or any group of people. Therefore, for me, an interrogation of the use of biometric technology in elections is necessary for two reasons: First, it provides an avenue to interrogate the social impact of technology in our society, and second it helps us to respond to Adam Banks' call on TCP and related fields to make technology access a major part of our inquiry.

Banks (2006) call is timely because the biometric, a digital technology, is being used to establish electoral integrity and this technology is increasingly becoming a global technology. I focus on discriminations inherent in the design of the biometric technology and question why the biometric disenfranchises specific kind of voters. Unquestionably, Adam Banks has been able to prove that the extent to which technologies can be used to upend race and discrimination, and Winner (1980) discusses how Moses' decision to make "the bridges over the parkways on Long Island, New York" (p. 123) extraordinary low "reflect Moses' social-class bias and racial prejudice" (p. ibid.). These pieces demonstrate how artifacts are political and can be used to discriminate against a specific group of people. In fact, Haas (2012) states aptly that "technology is integral to culture and always already cultural. Just as the rhetoric we compose can never be objective, neither can the technologies we design. Technologies are not neutral or objective—nor are the ways that we use them" (p. 288). In this regard, Angela Haas points out the need to examine the interstices between discrimination, race, rhetoric, and technology design since such an inquiry will have several implications for our field. For example, usability scholars will be pushed to "interrogate the extent to which all designers imagine users that mirror themselves—and calls into question the extent to which designers are capable of imagining users different from themselves" (Haas, 2012, p. 304). From a localization perspective, such an inquiry acknowledges and reinforces our humanities' values. I stand with all those scholars in our field who have argued that since technology has become ubiquitous; it is necessary to study how they rearticulate existing structures that dominate or oppress vulnerable populations (Haas, 2012; Jones, 2016a, b; Rose, 2016).

Such an inquiry should extend to the use of biometric technology in order to study the discriminatory tendencies embedded in it. This is important because it has been revealed by scholars that "biometric technology, in its current form, serves to categorize bodies with a dangerous discriminatory logic that cannot be touted as 'true' or 'objective'" (Murray, 2007, p. 359).

BIOMETRIC USE AND THE ENACTMENT OF DISCRIMINATION

Biometric production assumes that the human body cannot lie so it will be a gold mine for accurate information. The truth about human body is subjected to the lenses of the biometric device. Because the device captures the biological and physical features of individuals through the use of computers, scanners, and webcams, it is believed that the biometric collects accurate information about individual so that "when we say it is you, we know it is you" (stIghana, 2012). As a result of this trust that biometric captures accurately, the technology has become an integral part of our globalized, interconnected world; it is "used by government agencies and private industry to verify a person's identity, secure the nation's borders,...to restrict access to secure sites including buildings and computer networks" (Vacca, 2007, p. 3). The biometric device, to borrow from Katz (1992), embodies an ethic of expediency: rationality, objectivity, neutrality, efficiency, productivity, and speed.

The representation of the biometric as objective, neutral, accurate, and truthful technology, I have argued, is an attempt made by biometric scientists and biometric industries to make a legitimate claim to support the biometric ideology or the science project. Such representations, first, reinvigorate the popular notion that science is removed from cultural practices—this has proven to be a weapon used to produce and maintain the scientific project; and second, deny that biases and discriminatory tendencies can be programed into the biometric technology. Thus, to refer to biometric in light of scientific objectivity is an attempt to "accrue to it all the mute credibility and faith that" the technology represents (Barton & Barton, 2004, p. 237). While Steven Katz calls the very idea of technological objectivity a "fallacy," some feminist critics of science and technology have revealed that the claim to scientific or technological objectivity is untenable, which means that technology, which is an end goal of scientific theory, cannot be neutral or objective. As one such example, Judy Wajcman (1991) indicates that even the kinds of language used to discuss technology favors men. This intricate relationship, Wajcman argues, is not inherently biological but it is as a result of how gender has been culturally and historically constructed. As a matter of fact, Wajcman elaborately indicates that "technology is more than a set of physical objects or artifacts. It also fundamentally embodies culture or set of social relations made up of certain sorts of knowledge, beliefs, desires and practices" (p. 149). This means technology has a culture. Even though feminist arguments are made in regard to the relationship between gender and technology in Western societies, I make a case that it becomes relevant in a non-Western society because it provides avenues for scholars to interrogate the various ways that technologies that are adopted by non-Western societies carry the stigmata of Western cultural ideologies.

I must point out that even though it is not explicitly stated, localization experts in technical communication share the same concerns as feminists. To be sure, localization scholars in TC argue against a top-down approach to technology design. They contend that technologies break down because developers do not take into consideration the culture and nature of use in user contexts. Thus, designs do not reflect the ideology of users but instead, only make developer ideologies visible. To this end, localization experts have sought for ways to get designers to understand that users also have forms of knowledge which must be included in designs. As social justice and human-centered advocates, it is relevant to destabilize the notion that biometric is an accurate, objective, and a neutral technology that speaks for itself by reporting the various ways biometric fails. We need to fight for those who by no fault of theirs are rejected by the technology and are consequently refused to cast their ballots. Walton (2016) says, and I agree that "oppression disrespects the intrinsic worth of a person" (p. 412). Consider how one feels when the person has queued for several hours to vote only to be told at the voting table that he/she cannot vote because the biometric cannot read their fingers. Clearly, we need to fight injustice by conducting research that seeks to unravel the various ways technologies discriminate against others.

Decolonizing Biometric Use Through Social Justice and Localization

The conversation throughout this book, and especially in this chapter, indicates that technology use is situated in the intersection of localization and social justice. As I have said, localization is not about translation, but about the extent to which users demonstrate their knowledge of use by adopting and reconfiguring the purpose of technology to solve local problems. Localization theories advanced by technical communication scholars seek to encourage designers of technology to integrate users' ways of knowing into the design of technologies (Agboka, 2013; Johnson, 1998; Salvo, 2001; Sun, 2006), although these goals are yet to be fully realized (Spinuzzi, 2003). Localization aficionados have been able to make a profound argument for the creative way's users localize technologies. I agree that users have knowledge of use, that they have a voice and that they are not dummies but active agents who save technology. I think this message has received the needed acceptance. We have managed to argue that localization is beyond translation and that it must pay attention to local users and their logic. But how do we help users to communicate their agency to designers? How do we help users to make a case that their local forms of knowledge should be included in design processes? This for me is the next stage localization scholars should be thinking about: We have to find ways to help users make an active statement to designers for their ways of knowing to be included in design processes. We need to, as a matter of urgency, decolonize localization. Meaning, user-centered and localization theorists should not be so fixated on looking at ways to incorporate user knowledge into design situations; but instead look at the "limits" of technology, that is, the various ways technology fails and the nature of the failure. I am of the view that the transfer of technology to different parts of the world presents a challenge to localization. Godwin Agboka demonstrated this through his study of documentation accompanying aphrodisiacs imported to Ghana, while Huatong Sun's study of mobile technology use in two disparate contexts reveal the sense of disconnection that exists between the developers of the technology and the users.

I am of the view that localization researchers have created passive localization in that even though they make positive claims about the knowledge and agency users possess, they do not tell users how to enact their agency. More so, our research findings are hardly communicated to the users we advocate for. So, for instance, we argue in journals and books that users are powerful but how do we let them know they possess that power? How do we know they even have a voice which needs to be heard by designers? I believe that if we adopt failure narratives, we can more appropriately discuss how technologies adopted have limits and fail to address socio-physiognomic issues, such as worn-out fingerprints, or other forms of bodily deviance, in local contexts. That is how we can be more "user-sensitive" (Agboka, 2013, p. 42) in our approaches, and consequently, advance a social justice goal of empowering the disenfranchised.

We must ensure that "groups and individuals receive equal opportunities and are not marginalized and disenfranchised" (Jones, 2016a, p. 472) by technologies are designed, adopted and used. It is undeniable that localization and social justice thrive to empower the marginalized and the disenfranchised in the society. Colton and Holmes (2018) have argued that TC officials have only passively engaged with discussions about equality. The scholars recommend that a more active social justice concern supported by Ranciere's active theory will help users and TC officials to understand that "equality is something that any individual…can enact independent of a permissible institutional or governmental structure" (p. 12). This is the very reason why the rejection of voters by the biometric is not only a localization concern but also a social justice issue. Why? Because, social justice seeks to empower the marginalized or disenfranchised individuals in our societies or it thrives on equity, meaningful access and it provides individuals ability to "engage in the activities necessary to achieve what they want…" (Light & Luckin, 2008, p. 9). Those individuals who were rejected were willing to participate in the country's electoral process, but they were disenfranchised or disempowered. In the case of Ghana, voters who had queued for hours to elect their representatives were disallowed to participate because their fingers did not fit into the normative view of what a normal finger should be for it to be accepted by the biometric technology. Hence, such individuals were rendered powerless and had no means to exercise their franchise. Focusing on biometric use in elections, for instance, helps social justice advocates to discuss how humans are affected by technology design and use or how "technologies of measurement are affecting our ontological existence, our beingness and relations, and the essence of what makes us humans" (Beer, 2014, p. 334) and citizens. Our role as a more active social justice advocates is to emphasize the need to pay attention to how biometric technology fails. Hopefully, through our active and constant communication of biometric failures, biometric production companies will know about how the technologies they design discriminate against some people and act to salvage the situation.

The lived experiences of these users enable us to negotiate with biometric scientists to value human dignity and ensure that their technologies are designed to capture every individual in the society, including those with worn-out fingerprints. In essence, I am saying that biometric designers should "conscientiously employ...methods to engage purposively with issues of social justice and deliberately make more just and equitable design an end-goal" (Jones, 2016a, p. 477). To be sure, I am not making a case that the EC of Ghana purposefully sought to disenfranchise or discriminate against voters. In word and act, the biometric did save the electoral process of Ghana to an extent; instead, I seek to bring home the essential relationship between technology design, use and ethics in a resource-constrained context. I believe that discrimination "can occur without malicious intent, ill will, or even active engagement" (Jones, 2016a, p. 478). I am arguing that the biometric, a product of scientific culture, is not neutral as purported by biometric scientists and that it is possible that it has limits that work to discriminate against specific group of people as exemplified in Ghana's case. Based on the nature of biometric breakdowns, we may say that the biometric even reproduces the idea of the difference. Marginalization and the institutionalization of difference could be built into the design process of the biometric. It embodies disenfranchisement and difference because it is programed based on a limited scientific theory which wrongly assumes that the human body is stable across the globe (Magnet, 2011).

Decolonizing Biometric Use Through Concerns About Ethics

Who gets accepted or rejected by a design is a matter of ethics! My fear is that if we do not point out the various ways biometric technologies disempower and rid people of their human dignity, we may be reinforcing the ethics of expediency. By which I mean, we as technical communicators may overzealously make a case for the objective nature of the technology, as I pointed out in Chapter 5 while we neglect other moral issues. My concern is borne out of the fact that over two decades ago,

Steven Katz decried that a technology can be designed purposely to execute a nefarious act such as the extermination of a specific group of people. In "the ethic of expediency," he revealed how Just persuaded his superior to effect changes to railroad cars which will be used to exterminate "the Jews and other 'undesirables'" (p. 256). In this regard, Steven Katz points us to the ethical issues involved in the construction of technology (a van). He is concerned about the emphasis on expediency to the neglect of human rights and human dignity, chief concerns of social justice and localization.

I contend that technical communicators and social justice advocates will be complicit in unethical behavior if we glorify the technical prowess of the biometric or any technology and stand aloof to concerns that the biometric reject specific group of people who seek to exercise their civic rights. In "Design as advocacy," Rose indicates that technical communicators are better positioned to identify and question how technologies we use reinforce "existing inequalities into their designs and risk further disenfranchising already vulnerable populations" (p. 428). She warns, thus, that we cannot fail to account for vulnerable populations in technology design as doing so leads to the creation of technologies that produce, maintain, or reproduce inequality (p. 442). The goal of the technical communicator, thus, is to identify how technologies that are designed intentionally or inadvertently reinforce structures that maintain inequality. I indicated earlier that the bedrock of Ghana's economy is agriculture. It is estimated that about 50% of the country's Gross Domestic Product is derived from agriculture and the agricultural industry employs about 55% of the country's population is engaged in agriculture. In spite of the large role agriculture plays, farmers struggle to get access to good roads, storage facilities, good schools, good social amenities, good health systems, and advanced farming technologies. As a result, most of these farmers resort to the use of simple tools such as machete and hoes to cultivate their farmlands. The consequence is that most of them have worn-out fingerprints. Also, because they do not get enough support, most farmers wallow below the poverty line in Ghana, and thus, we can characterize the poor farmer as a vulnerable individual. Therefore to adopt a technology which could potentially reject people who have worn-out fingerprints is to inadvertently work to discriminate against them.

The ongoing conversation indicates that when technology becomes an ethos or a form of consciousness, there can be inherent dangers.

Therefore, whenever we think about the ethos of a technology we must as well think about the ethical implications of the design and use of the technology. I am not suggesting that biometric can be used in the same way as the Nazi's used vans to execute their unethical acts. Biometric "is all about measuring life, measuring uniqueness of the 'bio' and its identity." Which means it is "implicated in processes of categorization and classification which allow the (sub)division of the population into manageable groups according to their level of risk and identity profiles" (Beer, 2014, p. 331). We know from history that biological determinism, that is, classification of individuals based on biological behavior, has been used to categorize, exclude, and discriminate against specific groups of people. For example in *The Mismeasure of Man*, Stephen Jay Gould enumerates how worth was "assigned to individuals and groups by measuring intelligence as a single quantity" (p. 52). Two different approaches were used: craniology and the subjection of individuals to some kinds of psychological testing. According to Gould, determinists mostly dwell on scientific proofs which they claim are objective and free from social and political taint. These revelations by Gould are necessary because biometric scientists also make claims to the objectivity of science to support claims about the powerfulness of the biometric technology. More worrying is the fact that biometric technology uses measurement as the basis to generate its codes. This means we must interrogate "the very ontological foundations of biometrics, the assumptions this technology makes about who we are, and the ramifications of a technologically mediated approach to governing identity" (Beer, 2014, p. 330). It is also relevant to point out that even though biometric use is traced back to antiquity, modern forms of biometric, which were developed in the nineteenth century were purposely used to identify and control criminals and colonial subjects (Maguire, 2009, p. 10). Therefore since most of the countries which have adopted to use biometric to enhance elections have experienced colonialism at a point in its history, my fear is that biometric measurement is a recast of colonial rhetorics or recast of instances where measurement was/has been used to discriminate or categorize human beings.

Last but not least, the rejection of people with rough fingers, etc. during the voting process is indicative of the fact that biometric technologies suffer from "demographic failures"; that is, "they reliably fail to identify particular segments of the population" (Magnet, 2011, p. 5). To an extent, the rejected voters are cast as "monsters" or "illegible bodies" whose citizenship as Ghanaians and rights to vote were questioned. We are told that voters can only vote when they are identified by the technology. This means that one is disenfranchised if, through no fault of theirs, technology fails to identify them. Since it is a constitutional requirement for a Ghanaian who is 18 years and above and of sound mind to vote, that individual becomes a subject of concern because they could not exercise their constitutional rights. This indeed is a recipe to investigate the various ways that technologies work to infringe on the rights of individuals. The rejected voters are not only discriminated against or disenfranchised, they experience a form of silencing. Silence does not only refer to the inability to talk or make a speech, it can also mean "…not being included or represented, being ignored or delegitimized, not being valued, or being actively oppressed and marginalized" (Jones, 2016a, p. 478). The implication of Jones' statement is that an individual who is rejected by the biometric experiences a form of exclusion or marginalization. Such an individual cannot vote and thus is silenced; such an individual is excluded from performing his/her civic responsibility, and that individual is delegitimized.

We need to adopt a more critical approach to studying technology localization and social justice. I have revealed elsewhere (Dorpenyo, 2016) that one of the ways to critically analyze a technology to identify its localization and social justice implications is to "unblackbox" it. I use the term "unblackbox," a concept I borrow from Blake Scott (2003), to refer to the idea that technology is not neutral and that it is necessary to open it up in order to identify the various ways it operates among human lives. Technologies, such as the biometric, appear to be complex, neutral, and compact, but if we take our time to study it, to open it, we can understand the various ways using them can uncover issues of discrimination. Technology, such as the biometric, is a "black box" in the sense that it is a machinery that is "too complex," it operates on "complex set of commands" and it comes to users as a compact technology. Because of its complexity, we might think it is "discrete, coherent" (Scott, 2003, p. 22). A critic's job "is to reopen the black box…by treating it as a messy and dynamic enterprise…" (ibid.). We do not only have to account for the good of technology but also how their usage lead to a discussion of broader issues such as social justice, localization, power, ideology, discrimination and control. As Latour (1987) states, we need to "enter facts and machines" (p. 13) in order to "open the black boxes so that outsiders may have a glimpse at it" (p. 15).

Entering facts and machines, as exemplified in Chapter 5, becomes the way we can study arguments about technology—both implicit and explicit. We need to understand both implicit and explicit arguments about technology because "it is only after endless little bugs have been taken out, each bug" (p. 11) revealed that we can understand how such a technology can upend race, discrimination, and disenfranchisement. Taking out little bugs, by conducting research and reporting the moments the technology fails and the extent to which its failure hurts the dignity of human beings, can help us to intervene. This kind of study helps us to understand that using deterministic discourses to represent technology to users does not mean the technology will be successful. Ultimate success is judged by users' experience of that technology. If we are becoming dependent on technology, and if technology speaks about us, through verification and identification, then we need to pay attention to the various ways that technologies speak discrimination, race, marginalization, and disenfranchisement. We need to study the kind of subculture of outcasts that might be formed not by those who choose not to participate, but rather by those who are prevented from participating by the technology. Innocent bodies or individuals are blamed unjustly when they are rejected by the biometric when the fault, in fact, lies in the biases encoded into the design of the biometric technology.

References

Agboka, G. Y. (2013). Participatory localization: A social justice approach to navigating unenfranchised/disenfranchised cultural sites. *Technical Communication Quarterly, 22*(1), 28–49. https://doi.org/10.1080/10572252.2013.730966.

Banks, A. J. (2006). *Race, rhetoric, and technology: Searching for higher ground*. Mahwah, NJ and Urbana, IL: Lawrence Erlbaum; National Council of Teachers of English.

Barton, B. F., & Barton, M. S. (2004). Ideology and the map: Toward a postmodern visual design practice. In J. S. Johnson-Eilola & S. A. Selber (Eds.), *Central works in technical communication* (pp. 232–252). New York: Oxford University Press.

Beer, D. (2014). The biopolitics of biometrics: An interview with Btihaj Ajana. *Theory, Culture and Society, 31,* 329–336.

Bowdon, M. (2004). Technical communication and the role of the public intellectual: A community HIV-prevention case study. *Technical Communication Quarterly, 13*(3), 325–340. https://doi.org/10.1207/s15427625tcq1303_6.

Colton, J. S., & Holmes, S. (2018). A social justice theory of active equality for technical communication. *Journal of Technical Writing and Communication, 48*(1), 4–30.

Cushman, E., & Grabill, J. T. (2009). Writing theories/changing communities: Introduction. *Reflections: Writing, Service-Learning, and Community Literacy, 8*(3), 1–20.

Dorpenyo, I. K. (2016). *"Unblackboxing" technology through the rhetoric of technical communication: Biometric technology and Ghana's 2012 election*. Open Access Dissertation, Michigan Technological University.

Dubinsky, J. M. (2004). Guest editor's introduction. *Technical Communication Quarterly, 13*(3), 245–249.

Flower, L. (2008). *Community literacy and the rhetoric of public engagement*. Carbondale, IL: SIU Press.

Haas, A. M. (2012). Race, rhetoric, and technology: A case study of decolonial technical communication theory, methodology, and pedagogy. *Journal of Business and Technical Communication, 26*(3), 277–310.

Hobbis, S. K., & Hobbis, G. (2017). *Voter integrity, trust and the promise of digital technologies: Biometric voter registration in Solomon Islands*. Paper presented at the Anthropological Forum.

Johnson, R. (1998). *User centered technology: A rhetorical theory for computers and other mundane artifacts*. Albany: State University of New York.

Jones, N. N. (2016a). Narrative inquiry in human-centered design: Examining silence and voice to promote social justice in design scenarios. *Journal of Technical Writing and Communication, 46*(4), 471–492. https://doi.org/10.1177/0047281616653489.

Jones, N. N. (2016b). The technical communicator as advocate: Integrating a social justice approach in technical communication. *Journal of Technical Writing and Communication, 46*(3), 342–361. https://doi.org/10.1177/0047281616639472.

Katz, S. B. (1992). The ethic of expediency: Classical rhetoric, technology, and the Holocaust. *College English, 54*(3), 255–275.

Latour, B. (1987). *Science in action: How to follow scientists and engineers through society*. Cambridge, MA: Harvard University Press.

Light, A., & Luckin, R. (2008). *Designing for social justice: People, technology, learning*. Retrieved from https://www.nfer.ac.uk/publications/FUTL53/FUTL53.pdf.

Magnet, S. (2011). *When biometrics fail: Gender, race, and the technology of identity*. Durham, NC: Duke University Press.

Maguire, M. (2009). The birth of biometric security. *Anthropology Today, 25*(2), 9–14.

Murray, H. (2007). Monstrous play in negative spaces: Illegible bodies and the cultural construction of biometric technology. *The Communication Review, 10*, 347–365.

Preliminary Report on Monitoring the Right to Vote and Observing the 2012 Presidential and Parliamentary Elections in Ghana. (2012). Retrieved from Accra.

Rose, E. J. (2016). Design as advocacy: Using a human-centered approach to investigate the needs of vulnerable populations. *Journal of Technical Writing and Communication,* 46(4), 427–445. https://doi.org/10.1177/0047281616653494.

Rude, C. D. (2004). Toward an expanded concept of rhetorical delivery: The uses of reports in public policy debates. *Technical Communication Quarterly,* 13(3), 271–288. https://doi.org/10.1207/s15427625tcq1303_3.

Salvo, M. J. (2001). Ethics of engagement: User-centered design and rhetorical methodology. *Technical Communication Quarterly,* 10(3), 273–290.

Scott, B. (2003). *Risky rhetoric: AIDS and the cultural practices of HIV testing.* Carbondale: Southern Illinois University Press.

Simmons, W. M., & Grabill, J. T. (2007). Toward a civic rhetoric for technologically and scientifically complex places: Invention, performance, and participation. *College Composition and Communication,* 58(3), 419–448.

Spinuzzi, C. (2003). *Tracing genres through organizations: A sociocultural approach to information design* (Vol. 1). Cambridge, MA: MIT Press.

stlghana. (2012). *STL Ghana—Biometric system voter registration process.*

Sun, H. (2006). The triumph of users: Achieving cultural usability goals with user localization. *Technical Communication Quarterly,* 15(4), 457–481. https://doi.org/10.1207/s15427625tcq1504_3.

Sun, H. (2012). *Cross-cultural technology design: Creating culture-sensitive technology for local users.* New York: Oxford University Press.

Vacca, J. R. (2007). *Biometric technologies and verification systems.* Burlington, MA: Butterworth-Heinemann.

Wajcman, J. (1991). *Feminism confronts technology.* Pennsylvania: The Pennsylvania State University Press.

Walton, R. (2016). Supporting human dignity and human rights: A call to adopt the first principle of human-centered design. *Journal of Technical Writing and Communication,* 46(4), 402–426. https://doi.org/10.1177/0047281616653496.

Winner, L. (1980). Do artifacts have politics? *Daedalus,* 109(1), 121–136.

CHAPTER 9

Conclusion: Participatory User Localization

Technology transfer has received some attention in the field of technical communication. This interest in finding the relationship between technical communication and technology transfer is well expressed in the special issue organized by *Technical Communication Quarterly* in 2006. The editor of this special issue, Nancy Coppola, indicated that even though a lot of researchers have looked into technology transfer from various angles, it was "Doheny-Farina who set the field for us" (Coppola, 2006, p. 287) when he published *Rhetoric, Innovation, Technology: Case Studies of Technical Communication in Technology Transfer*. In this book, Doheny-Farina explored the rhetorical nature of the phenomenon of technological transfer and in the process "uncovers some of the rhetorical barriers to successful technological transfers" (1992, p. ix). He argued for the study of every aspect of technological transfers from a rhetorical and socially constructed communication model.

It becomes relevant to study technology transfer in international context because the technologies that are transferred "develop complex relationships to each cultural-rhetorical tradition across the globe" (Thatcher, 2006, p. 383). Ghana's case becomes an avenue to interrogate an aspect of how a technology which is adopted develops this complex relationship Barry Thatcher discusses. He identifies that the relationship could be that of fit and reciprocity, that is, the extent to which the technology correlates with the cultural situations of the target context (fit) and how the fit situation evolves as the culture and technology evolves. By these two principles, Thatcher is able to establish that there is a relationship between technology and culture at every instance. He admits, though, that it has not been an easy task for international technology transfer scholars and scholars of rhetoric and technology studies to figure out what the exact relationship entails. Thatcher (2006) proposes five critical points that can lead to a useful study of how technical communicators can capture this relationship in technology transfer situations: (1) the relationship that exists between rhetoric, culture, and the purposes of technical communication in technology transfer; (2) how technical communication reflects the rhetorical tradition from which the technical communicators come; (3) the kinds of adaptations needed to meet the rhetorical needs of the adopted contexts; (4) how larger contexts encourage local, organizational or personal rhetorical strategies in the integration process; and (5) how the rhetoric surrounding the technology will speak about what is naturalized (p. 385). Ultimately, such a study should move toward how to "unmask assumptions from both cultures about the roles of writing in creating or maintaining certain types of rhetorical, social, organizational, and cultural interactions in new technology contexts" (ibid.).

Taking a cue from the above, this book contributes to conversations about localization, a core aspect of technology transfer, by studying Ghana's use of biometric technology in its electoral process. It proposes the use of decolonial methodologies in studying technology use in non-Western, colonized contexts. Through the analysis of interviews and technical genres, I explore and advance three localization strategies: linguistic localization, user-heuristic experience localization, and subversive localization. I also identify a localization cycle (Fig. 9.1).

My argument is that users are creative and innovative, and they devise different strategies to help them integrate technology into their local contexts. Designers may have to understand that users subvert and reconfigure the purposes of the technology they design and instill flexibility into their design. This concluding chapter elaborates on my contributions to localization, to the study of election technologies in a global era, technology transfer, international/intercultural technical communication and to professional and technical communication.

I cannot overgeneralize and say that the user strategies adopted by Ghanaians to integrate the biometric into the electoral system represent the various user strategies in every part of the world where biometric is used to enhance electoral processes, but the case indicates that for a technology adopted to be successful, there must be the active participation

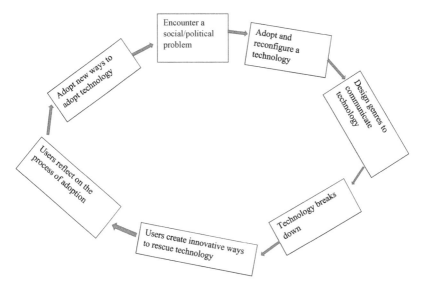

Fig. 9.1 Localization cycle

of users. Hence, the title of this chapter: participatory user localization. This concept blends ideas from Godwin Agboka's (2013) "participatory localization" and Huatong Sun's "user localization." In the first, Agboka proposes that special attention should be paid to social and political conditions in the user's context. Agboka does not focus on isolated individual user participation, but instead, "user-in-community" participation in design and adoption processes. In Huatong Sun's conception, user localization focuses on various strategies users adopt to integrate technology into their daily lives. As she explains further, user localization is both a product and a process. As a product, user localization indicates the intricate relationship "with a locale" (Sun, 2009, p. 258) and the cultural and social factors that emanate from that culture; as a process, localization indicates user interactions that enable the integration of a technology into a user's culture. On another level, user localization is a work of articulation. In this regard, localization links instrumental processes of technology integration with subjective user experiences. The implication is that localization should emphasize community engagement of users, social, political, and environmental factors. When this happens, technology designers will recognize user experience than a mere communication of instrumental features of a technology. In Chapter 5, I revealed that a gap exists between what we say about technology and how that technology functions. To properly represent a technology is to tell stories about how community users are able to integrate that technology into their society.

This means localization is not complete unless it involves an active participation of users in a local context. For example, after the 2012 elections, the opposition NPP filed a petition to the Supreme Court of Ghana contesting the results. The eight-month-long legal tussle revealed several incongruities in the electoral process of Ghana. More pertinently, the Supreme Court revealed that more has to be done to integrate the biometric technology into the electoral process of the country. The Judges unanimously agreed that reforms are pertinent to the success of subsequent elections and suggested, inter alia, that "the biometric device system must be streamlined to avoid breakdowns…" (Baneseh, 2015, p. 167). In fact, most of the EC officials I interviewed accepted that the petition filed did a lot of good to the electoral process and it contributed to the successful use of the biometric in the 2016 elections. Mr. Manu, as an example, indicated that "out of the petition came the recommendations from political parties, stakeholders, civil society groups and all those things and these are the measures that we actually took because if you look at the competence of the election officials, they were some were saying twenty-seven zero and other things." This means that users (political parties, civil organizations, the Supreme Court, and other stakeholders) had a stake to play in the successful integration of the biometric. When these civil organizations were sidelined during the initial implementation stages, the country was almost plunged into chaos but when they actively participated after the 2012 elections, the country saw success in subsequent elections. 2016 elections will go down as one of the most successful elections in the country's history; thanks to the active participation of users in the electoral process. The success of biometric use cannot solely be attributed to election technology; we can acknowledge the dynamic interaction between various users and stakeholders at various stages.

The example above indicates that the active participation of users in a community is relevant for successful integration of technology. This is consistent with Agboka's (2013) call for "user-in-community involvement and participation" (p. 42) in technology design processes. For me, users should not only be involved in the design process but the integration process as well. The idea of community participation in localization processes is more relevant because Ghana is classified as a Collectivist society. In a Collectivist society, individuals view themselves as dependent on others, and thus, define themselves as part of a group. In these collective societies, "communication patterns emphasize interpersonal relationships, social hierarchy, social leveraging, group identities, close personal space, and writer-friendly patterns" (Thatcher, 2011, p. 47) and "individuals feel a deep personal involvement with each other" (Sun, 2012, p. 213). Therefore, if Godwin Agboka argues that users in a community should be actively involved in technological design processes, he is stating a core aspect of localization. Though it is difficult to have every user actively participate, those willing to participate must be given the opportunity to do so—in essence, the process must be "mutual and participatory" (Agboka, 2013, p. 44). The process can also be designed around Johnson's (1998) participatory model: "The rhetorical situation involving users, designers, and artifacts should interact in a negotiated manner so that technological development, dissemination, and use are accomplished through an egalitarian process that has its end in the user" (p. 85).

Mutual user participation is necessary because technology use is situated in user's complex context. Since technology use is situated in a users' context, the design of a technology should also be situated in user context and must be designed to support concrete user actions. If concrete user contexts are ignored, or if localization is decontextualized, users in a community will not properly appreciate the technology which has been adopted or integrated. So, for instance, even though biometric is enhanced with security features, designers may have to design different biometric technologies which support electoral needs. As I indicated in Chapter 4, users revealed that among other things, Ghana's electoral process sees an increasing number of minors and foreigners attempting to vote but the biometric is not designed to identify such people. Is it possible for biometric designers to start thinking about the features they need to add to biometric technologies? This means that localization should be an ongoing conversation between users and designers. It also means designers should consider the social interactions that surround the practices in which biometric technology is introduced.

More so, my analysis indicates that localization is a reflective and social learning process; hence designers must understand the various learning processes of users they target. Perhaps, more should be done to accommodate errors by users or reduce complexities that enhance error making. Social learning, thus, presents an interesting question about "how people gain the competences that are needed to participate in a social activity and how they become proficient actors within given systems of activity?" (Tuomi, 2005, p. 30). As indicated by one interviewer when asked how Ghana eventually used the biometric technology successfully in 2016, he responded, "...So learning from the challenges of 2012." In Ghana's case, this process of social learning involved a complex articulation of: making mistakes, learning from past errors, questioning the validity of a technology, and even adopting heuristic approaches to resolve errors. These varied experiences exemplify Lave and Wenger's (1991) concept of "communities of practice" which suggests that tyros become experts through processes of socialization and articulation. Through these processes of socialization, they are transformed from neophytes into expert members of a community. As I have indicated severally, Ghanaians transformed from being non-expert users of biometric technology, in which case a lot of break downs were encountered, to expert users, where little errors or break downs were recorded.

More importantly, the Ghanaian examples exemplify Lave and Wenger's argument that "learning is an integral and inseparable aspect of social practice" (p. 31). "Social practice" as a concept, has received much attention by literacy scholars and when looked at carefully could help advance conversations about localization. "Social practice" as discussed in literacy studies emphasizes the understanding of literacy practices within the social and cultural contexts in which they are practiced (Street & Lefstein, 2007). David Barton and Mary Hamilton, as an example, presented a theory of literacy practices which identified that "social practice" refers "primarily [to] something people do; it is an activity located in the space between thought and text" (Barton & Hamilton, 1998). Specifically, they reveal that "practices".

- include people's awareness of literacy, constructions of literacy, and discourse of literacy at the same time practices are the social processes which connect people with one another"
- "are shaped by social rules which regulate the use and distribution of texts, prescribing who may produce and have access to them" (Barton & Hamilton, 1998)
- straddle the distinction between individual and social worlds, and literacy practices are more fully understood as existing in the

relationships between people, within groups and communities, rather than a set of properties residing in individuals" (Barton & Hamilton, 1998).

We therefore must redefine localization in ways that perceive of the user as engaged in "social practice" (Tuomi, 2005, p. 27); meaning that the biometric must be perceived of as a tool which has a social dimension and that its design must be tied to a specific social activity in a context. As I said earlier, biometric could be designed for election purposes. This may come with some expense, but it will be worthwhile. If the product which has been adopted cannot resolve all the problems that hamper successful elections, or if users struggle to learn and use the functionalities to support electoral integrity, the biometric may die away. It also means social rules that regulate the biometric use must be looked at. For instance, when the biometric technology was adopted in 2012, the parliament of Ghana enacted a law which ensured that only the biometric could be used as a means of authentication and verification, a law which was captured in simple terms as "No Verification, No Vote." After a period, the country realized that the law was not going to be very effective, so some people started acting in ways that contravened the stipulated laws managing biometric use in elections. In subsequent years, this law has been revised to accommodate the needs of the Ghanaian context; this is an indication that users transform in their learning processes.

If localization centers on user strategies/tactics to salvage technological breakdowns as I have indicated here, and Sun emphasized, then we can fall on rhetorical theories to reconceptualize localization. Far too often, scholars have sought to explain localization from a cultural perspective. The consequence, as these scholars reveal, is the narrow conceptualization of localization. de Certeau makes us understand that tactics can operate on rhetorical terms. As de Certeau reveals, "tactics" dwells on propitious timing (Kairos) and situational cunning (metis). I have indicated how Kairos and metis are relevant to localization through Ghana's case. I argued in Chapter 5 that the adoption of the biometric was kairotic as several stakeholders in Ghana lost confidence in the Electoral Commission of Ghana so the adoption of the biometric provided a timely intervention. Subsequently, in Chapter 7, I looked at how Ghanaian users exhibited their practical knowledge by redesigning the user manual which accompanied the biometric technology. I emphasized how they subverted official knowledge by dwelling on their cunning intelligence in their bid to solve complex technological problems. Sun (2012) also indicated how users devised different tactics to integrate technology into their local lives. These tactics/strategies employed by users indicate that they "use local tactics to make do with technology" (Kimball, 2006, p. 75). Thus, localization should ensure:

- Flexibility into design
- Relationship between designers and users
- Critical attention to relationship between artifacts, users, and environment
- Examination of user knowledge in a global context
- Design process should be extended to sites-in-use.

IMPLICATIONS FOR TECHNICAL COMMUNICATION

I have identified that for technical communication scholarship to fully achieve localization and social justice goals, we need to adopt a more active stance by critically tracing and analyzing discourses surrounding specific technologies, conduct field study to understand user experience of the technology and finally move to intervene. The time is ripe for TC scholars and practitioners to "challenge the promises of the biometric technology, to understand its logic, and to question its motivation. It insists that we pay attention to the positive and negative spaces of biometric technology's archives that we define it by the bodies it denies as much as by the bodies it promotes" (Murray, 2007, p. 360). When we take on this challenge, then, our core focus on humanity will be realized; our call to advance human dignity and human rights will be much appreciated by the users' who are discriminated. Localization will also take a more critical approach. Localization scholars will begin to care not only about how to include user knowledge into technology design; but also, how to talk back at biometric and technology designers.

Also, technical communicators are traditionally identified as accommodators of technology to users. In this regard, we are perceived as heroes who help users understand and use technology. But now that users can devise strategies to accommodate or integrate technology into their local contexts, what becomes of technical communication practice, research and pedagogy? These are days that we need to live by Savage's (2004) advice that we should train technical communicators as sophists, "not mastering but negotiating continually shifting technologies, institutions, discourses, and cultures" (p. 189). We also need to take seriously arguments that technical communication should go beyond the study of organization. With the inevitable growth in technology transfer and global flows, we need to train "savvy, flexible, tactical technical communicators" (Kimball, 2006, p. 83). As Kitalong (2000) points out, it is necessary to pay attention to media representation because: (1) it helps us to notice and acknowledge how complex processes of technologies are portrayed in magical terms; (2) it broadens the scope of technical communication scholarship to include media analysis. In this regard, the traditional notion that technical documents are the only means to which users get to know or understand technology is challenged. Therefore, scholars will not only turn attention to such documents as instruction manuals, warning labels, and training guides but also media representations; and (3) while technical communicators consume media discourses about creators, they also create technical discourses about technology. In essence, technical communicators should be encouraged to look beyond the traditional organization to uncharted paths such as technology adoption process in decolonial contexts.

My conclusion stresses the need for research to influence practice and pedagogy. I also stress the need for technical communicators to engage in civic activities that can enable them to identify and intervene to solve local and global problems. My main take in this research is that technical documents do not merely "accommodate technology to users" (Dobrin, 2004); Or that when we are engaged with an analysis of a document, we are not merely dealing with the texts in the document; rather, we are dealing with broader historical, social, political, power, economic, legal and ideological issues that circumscribe the document. We have to, therefore, open up to the various ways that texts open us to these broader issues. It is by opening up that we will work toward achieving the goals of unblackboxing I discussed in Chapter 8.

The important point to note is that decolonial methodology enables scholars to shift attention from studying "large cultures" to small subcultures, such as elections and how individuals negotiate their situatedness in their quest to claim agency. For instance, by studying the electoral culture of Ghana, we realize that the country has adopted several means in its quest to conduct free fair and incontrovertible elections. The structures in place during the 1992 elections have gone through drastic changes. Thus, we see a sense of dynamism in the Ghanaian electoral process. By studying electoral culture and small cultures, we can perceive culture as dynamic, ongoing process subject to articulation and rearticulation. Electoral culture in Ghana is not stable—it is in a state of flux. As one of the Chief Justices commented after the historical petition and as it was reiterated by the presidential candidate of the NPP, thus, "elections in Ghana will never be the same" (Group, 2014). Hence, to ascribe a cultural notion to Ghana is to prevent us from identifying the various processes that subcultures in Ghana are constantly undergoing changes. Elections in Ghana is almost always in the "process of adjusting and becoming" (Mao, 2005, p. 429).

My work demonstrates that there are similar ways of representing technology in both non-Western and Western societies. Or we can say that there are universal assumptions about technologies in both Western and non-Western societies. Under "biometric as the speaker of gendered patterns," I identified how the representation of the biometric technology as a technology that ensures "one man one vote," reinforce ideologies that inscribe masculine tendencies to technologies. As I indicated in Chapter 5, Western feminist scholars, such as Wajcman, Durack, and Koerber, have discussed the gendered nature of technologies. In as much as we would want to think that the representation of the biometric technology is shaped by Western forms of thought and logic, we would also want to consider or study non-Western forms of representing technology in their own terms. That is, we must acknowledge what Haraway terms "situated knowledge" about the biometric technology. In this direction, what is necessary to consider is Kenneth Pike's etic (insider) and emic (outsider) influences. The emic approach enables us to acknowledge the importance of local-based knowledge. For instance, Kofi Ghana, the expert user in the video I analyzed, represents a local authority who makes knowledge about the biometric in his bid to educate his fellow electorates. Meanwhile, etic approach enables us to recognize that there are similar universal forms of knowledge making. Thatcher (2010) stresses the importance of combining both etic and emic forms. Etic, he states, hinges the notion that there are "common human thresholds," whereas emic emphasizes the ways cultures conceptualize and operationalize these "common human thresholds" in unique and dynamic ways (p. 13). As Mao (2003) indicates, etic and emic approaches direct our attention "toward materials and conditions that are native to these traditions and so that appropriate frames and language can be developed to deal with differences as well as similarities between different traditions" (p. 418).

What we make of etic and emic frames is that it is relevant to pay attention to local situations and external influences. As I have indicated in Chapters 5 and 7, the various documents I analyze enable an understanding of the local situation that called for the introduction of the biometric technology. Through the analysis of documents, we identify that Ghanaians understand the need to persuade users of technology; they understand that we can use technology to solve complex problems and they also understand that rules and structures must guide technology use. Thus, as professional communicators interested in technology studies, it is necessary to pay attention to local material conditions that call for technology adoption and use. We need to analyze those conditions on their own terms before we can apply Western theories because "while non-Western… practices should be studied on their own terms, it does not mean that they remain on their own terms forever. In fact they cannot…because the 'here and now' will always be exerting pressure on these emic accounts, and the 'etic' moment will always be intruding on them" (Mao, 2003, p. 418).

To technical communicators studying technology in international contexts, these etic and emic frames will not only improve "the logics, fairness, and validity of the research" as Thatcher (2010, p. 13) indicates, it will yield "reflective encounters" (Mao, 2003, p. 418). Reflective encounters enable "creative understanding of different rhetorical traditions" (ibid.). Although Mao was addressing rhetors, I find this call for a reflective encounter between emic and etic frames a worthy step. The encounter will enable fruitful cohabitation between traditions and ways of knowing. More importantly, reflective encounters "renounce domination, adjudication, and assimilation, and they nurture tolerance, vagueness, and heteroglossia" (ibid., p. 418). In my case, I used Western rhetorical and cultural theories to study a non-Western context not because I feel Western forms of knowing are superior and more worthy than non-Western concepts but because I acknowledge the dialogic relationship between the local and the global. In any case, the biometric technology is Western technology and Ghanaians practice a Western form of democracy. Thus, it will not be out of place to use a Western lens to understand a non-Western situation. What is necessary for international technical communication scholars is a sense of reflection on the encounter between the theories they use and how they represent the material conditions they are studying because "…being local is not only about physical and temporal locality; it is also about the perspectives, the language ideologies, the local ways of knowing, through which language is viewed. The local should not be confused with the small, the traditional, the immutable, since it is also about change, movement and the production of space, the ways in which language practices… create the space in which they happen" (Pennycook, 2010).

More so, we will begin to reconsider received knowledge about the purposes and uses of technical documentation. As I demonstrated in Chapter 7, the user manual did not only aid Data Clerks to use the biometric machine, its analysis reveals how users can subvert technologies that are adopted and used. The user manual becomes an avenue to discuss concepts such as ideology, identity construction, subjectivity, law, local logics, and the various ways that technologies control us. Further, the study reveals the various ways that "rhetoric can function as biopower, wrapped up in larger power alignments that shape bodies and forms of embodiment" (Scott, 2003, p. 229). I point out in Chapter 5 that it is necessary to study rhetoric around technology, but it is not enough to limit such a study to just the rhetorical productions. It is paramount to understand how such rhetorical productions lead to a discussion of broader cultural entanglements. I demonstrate this by operationalizing Foucault's biopolitics and disciplinary technology and Haraway's cyborg theories as windows to understand the biometric as a tool that works to articulate such issues as gender and technology; technology and politics; the increasingly blurring boundaries between technologies and humans; technology and security; technology and ideology; and technology and identity construction. I also touched briefly on the question of citizenship and voting.

Subsequently, as my work indicates, it is necessary for technical communicators to become political. I do not mean to say that all technical communicators should aspire to become senators, governors, parliamentarians, or presidents. What I mean is we should "examine the way […] present conditions come about and are maintained for the purpose of understanding, critiquing, and changing them" (Blyler, 1998, p. 36). In Chapter 8, I dwell on Marion Young's concept of justice to enable an understanding of why elections in Ghana experience such malpractices as over voting. I point out that to answer this question, we have to work to study the various institutions charged to handle elections in Ghana. We also have to interrogate activities of political parties' and stakeholders and how their activities disrupt fair elections. In essence, we have to aspire to intervene in local and global situations. One of the ways to do this is to cultivate an interest in the public sphere. To an extent, we must aspire to become what Johnson (1998) refers to as "participatory citizens." A participatory citizen acknowledges that he/she is "a member of a larger social context" (p. 62). If we consider ourselves as citizens who have to participate in the affairs of a larger society, then, we will be able to carry the knowledge we acquire into the political arena, the space created for decision-making. In the political arena, Johnson states, we are invited to participate and we, in return, should respond to the call of participation by contributing to the common "good of the community." Our engagement must not necessarily be aggressive as we see in our daily political debates and contests. It can be peaceful, less provocative, metistic and nonviolent. My project demonstrates that as technical communicators, we can in our small ways help find answers to the problems we encounter in our societies (p. 66). We need to apply our "specialized knowledge" in service of community (Bowdon, 2004, p. 329). It is when we engage in civic activities and write about them that we can respond to the "political turn" of writing. We become public intellectuals.

My work does not only instigate political activism. It also contributes to discussions about civic engagement (Bowdon, 2004; Cushman & Grabill, 2009; Dubinsky, 2004; Flower, 2008; Grabill, 2001; Grabill & Simmons, 1998; Rude, 2004; Simmons & Grabill, 2007) in the sense that it engages in issues relating to the politics of Ghana. My major aim is to help Ghanaian electorates to think about the role they want technology to play in the electoral process and integrate the biometric properly so that what occurred after the 2012 election will not happen again (I am making reference to the electoral petition). I have expressed my civic responsibility by discussing issues that can lead to ways that technology can be integrated into the electoral process of Ghana. I discuss broader issues such as deliberations in Ghana in regard to elections. I also interrogate the issues that led to the adoption of the biometric technology. More examples of the relationship between technical communication and civic engagement can be found in the Fall 2004 special issue of *Technical Communication Quarterly* devoted to civic engagement and technical communication. In this issue, Dubinsky (2004), the Guest Editor, stresses the age-long argument that education should prepare students for civic engagement, meaning students should be given the tools to "exercise political power by pursuing goals concerned with 'human life and conduct'" (p. 245). The scholars who contributed to this issue identified the varied roles technical communicators play in the public sphere. One of such scholars is Bowdon. In the article "Technical Communication and the Role of the Public Intellectual," Bowdon (2004) tells a story of her involvement with a local AIDS prevention program to exemplify how technical communicators can act as public intellectuals. According to Bowdon, with our understanding of language as an ideological paradigm and genres as socially and politically situated, we can contribute responsibly in the public sphere by bringing our "specialized knowledge to serve as a liaison" (p. 329) among stakeholders in our communities. Though she was only invited to this AIDS project as an editor of technical documents, she realized that she was involved with something larger than "just editing." Through the narration of her participation in this project, she helps us to understand that as public intellectuals, the onus lies on us to: "make our work part of the public sphere;" "recognize the democratic functions of our work as educators;" "create positive changes in our communities by recognizing kairotic moments for intervention;" and "recognize our own situatedness within work contexts" (pp. 325–326). It is necessary to expose students and ourselves to ways that they can initiate local action to effect changes in their communities (p. 326).

Thus, as I have indicated elsewhere (Bowen et al., 2014), and as I have demonstrated in this book, civic engagement or the "public turn" of rhetoric and technical communication moves rhetors from their ivory towers on university campuses to communities around them. "Public turn" of rhetoric helps us to bridge the age-old binary between theory and practice. We bring theory to practice and practice to theory in service of our communities; we become agents of change. The "public turn" of rhetoric is not only interested in academic conversations, but it also gives rhetors the tools to apply practical judgment (phronesis) to solving community problems. We are ambassadors of change. Reflection and social action are the core tenets of community literacy and rhetoric. The public turn, for me, becomes a moment of engagement with the community around us in order to explore issues of concern and a moment of helping one another in order to solve complex problems. Rhetoric becomes the medium of exploration and a tool for inquiry.

I have demonstrated what the scholars have discussed not only through this project but through my teaching. As someone who is interested in civic engagement, I have found it necessary to create a dialog between my research and my teaching. I have not left my students out of the projects I engage in. Mostly, I tell stories about my research and how that demonstrates how technical communicators can be effective in their societies. For instance, my ENGH 388 Professional and Technical Writing students are engaged in the preliminary process of revising the user manual that accompanied the biometric technology. Dubbed "the big project," this assignment requires students to revise and design a user manual that accompanied the biometric verification device that was adopted by Ghana in 2012 for its presidential and parliamentary election.

The assignment extends civic engagement beyond local contexts to international and global contexts. As Walton (2013) and other scholars have identified, even though our societies and workplaces have become a globalized environment, most civic engagement activities in the academy are limited to local communities in which universities are located. Walton (2013) argues, and I agree, that the appropriate way to prepare our students to embrace the kind of globalized environment that we live in is to "seek civic engagement opportunities within not only local contexts, but global as well" (p. 148). Through the "Big project," students come to appreciate that globalization is about "cultural blending, hybridization, glocalization, and cross-border flow of rhetorical and cultural patterns" (Thatcher, 2011, p. 42). They learn how to appreciate the history, culture, politics, and rhetorical patterns of societies that are miles apart from their own. For example, while they work on the project, they ask questions about the currency of Ghana, political system, how to design a budget for an audience that is different from theirs. I appreciate and admire how they struggle to convert dollars to Ghana cedis, and how they eagerly read about the Electoral Commission of Ghana. More importantly, they learn how to analyze audience, collaborate with peers from diverse disciplines, and how to present information to diverse populations. In this regard, the students in my class (I hope) will "develop a more sophisticated knowledge of their own communication practices" (Brady & José, 2009, p. 41) and then go on to understand globalization to mean a process that can help them to acquire new ways of learning and doing things.

After analysis, they write a report on the state of the manual. This report assignment helps them to identify sections of the user manual which needs to be revised. I follow the report assignment with the proposal writing process where they persuade the Commission to release funds for a revision of the manual. They proceed with manual revision when I report to them that the Commission has accepted that they should revise the manual. While working on these projects, I assign readings that will enable them to reflect on cultural encounters and the necessity to value rhetorical and cultural patterns of different contexts.

By working with students to revise the user manual that accompanied the biometric technology, I adhere to Carolyn Miller's claim that we should teach a kind of "enculturation" that will enable students to understand how to belong to a community. Belonging to a community is a quintessence of technical communication. In a foreword to an edited collection of articles in *Reshaping Technical Communication: New Directions and Challenges for the 21st Century*, Janice Redish states that "community is going to be a major theme of this new century" (Mirel & Spilka, 2002). As Miller suggests, if we want students to take social action, we should involve the virtue of practical wisdom or *phronesis* (the ability to reason about the ends and not the means). "*Phronesis* enables a person to deliberate about the good rather than the expedient and, as such, to act in the political sphere rather than in the sphere of work" (Sullivan, 2004). Social action, based on *phronesis* enables students to become good citizens. The students must leave technical communication classes ready and willing to take socially responsible actions.

Implications for Biometric or Technology Designers

The biometric did not breakdown because Ghanaian users are dummies and cannot figure technology out (in fact the various theoretical constructions in this chapter indicate that Ghanaian users are smart users); in word and deed, the technology broke down because designers did not take into consideration the space, place, and location where technology is being used. The use of the biometric technology in elections is a complex and dynamic conversation with its surrounding social, cultural, technological, historical, and political condition. It talks about how locale forms an important part of technology adoption. To understand this complex relationship, we need to understand what spatiality means to technology use. How does Ghana's case help examine such terms as space, place and location in localization processes? Sun (2009) alludes to two defining arguments about dimensions of space and place in technology use: "space offers a concrete way to measure the distance that the convenience of mobile technology has brought us" (p. 246). Place which was introduced by Dourish, "refers to the way that social understanding conveys an appropriate behavioral framing for an environment" (p. 247).

Place as conveyed by Dourish, has a practical dimension to it and it helps to acknowledge the social dimension of a technology rather than "the instrumental convenience" (p. 247).

Sun argues that the three concepts of space, place, and locale enable us to concentrate on the relationship between the physical features of a technology and the local situations that determine the adoption of the specific technology. Though physical features of a technology are necessary, it is untenable to just focus on those physical features. It is very important to "understand what the changing spatiality means to our daily communication" (p. 247). This point is exemplified in Chapter 5 of this book where I looked at the rhetorical moves adopted by the EC to persuade the Ghanaian voter on the efficacy of the biometric technology. I identified how a narrow focus on communicating the instrumental features led both EC officials and lawmakers to ignore how the technology will perform when it is deployed to polling stations. As Chapters 6 and 8 revealed, the biometric technology broke down at the user level. In this regard, biometric use in elections is "an issue of interactions and encounters framed in" (p. 247) a rhetorical cultural context, and it presents possibilities for examining a "full spectrum of experience" (p. 247). More so, local situations determine how a technology such as a biometric technology is adopted, used, and consumed in a context as people use the technology to enhance their electoral systems. The implication is that we should see biometric technology beyond the physical and instrumental features to the interactions with surrounding context.

References

Agboka, G. Y. (2013). Participatory localization: A social justice approach to navigating unenfranchised/disenfranchised cultural sites. *Technical Communication Quarterly, 22*(1), 28–49. https://doi.org/10.1080/10572252.2013.730966.

Baneseh, M. A. (2015). *Pink sheet: The story of Ghana's presidential election as told in the Daily Graphic*. Accra: G-Pak Limited.

Barton, D., & Hamilton, M. (1998). *Local literacies: Reading and writing in one community*. Oxford: Psychology Press.

Blyler, N. (1998). Taking a political turn: The critical perspective and research in professional communication. *Technical Communication Quarterly, 7*(1), 33–52.

Bowdon, M. (2004). Technical communication and the role of the public intellectual: A community HIV-prevention case study. *Technical Communication Quarterly, 13*(3), 325–340. https://doi.org/10.1207/s15427625tcq1303_6.

Bowen, L. M., Arko, K., Beatty, J., Delaney, C., Dorpenyo, I., Moeller, L., ... Velat, J. (2014). Community engagement in a graduate-level community literacy course. *Community Literacy Journal, 9*(1), 18–38.

Brady, A., & José, L. (2009). Writing for an international audience in a US technical communication classroom: Developing competences to communicate knowledge across cultures. *Nordic Journal of English Studies, 8*(1), 41–60.

Coppola, N. W. (2006). Guest editor's introduction: Communication in technology transfer and diffusion: Defining the field. *Technical Communication Quarterly, 15*(3), 285–292.

Cushman, E., & Grabill, J. T. (2009). Writing theories/changing communities: Introduction. *Reflections: Writing, Service-Learning, and Community Literacy, 8*(3), 1–20.

Dobrin, D. (2004). What's technical about technical writing. In J. S. Johnson-Eilola & S. A. Selber (Eds.), *Central works in technical communication* (pp. 107–123). New York: Oxford University Press.

Doheny-Farina, S. (1992). *Rhetoric, innovation, technology: Case studies of technical communication in technology transfers*. Cambridge, MA: MIT Press.

Dubinsky, J. M. (2004). Guest editor's introduction. *Technical Communication Quarterly, 13*(3), 245–249.

Flower, L. (2008). *Community literacy and the rhetoric of public engagement.* Carbondale: SIU Press.

Grabill, J. T. (2001). *Community literacy programs and the politics of change.* Albany: State University of New York Press.

Grabill, J. T., & Simmons, W. M. (1998). Toward a critical rhetoric of risk communication: Producing citizens and the role of technical communicators. *Technical Communication Quarterly, 7*(4), 415–441. https://doi.org/10.1080/10572259809364640.

Group, J. (Producer). (2014, October 11). *Nana Akufo-Addo's Concession Speech on the Supreme Court Ruling on the 2012 Presidential Petition.*

Johnson, R. (1998). *User centered technology: A rhetorical theory for computers and other mundane artifacts*. Albany: State University of New York.

Kimball, M. A. (2006). Cars, culture, and tactical technical communication. *Technical Communication Quarterly, 15*(1), 67–86.

Kitalong, K. S. (2000). "You will": Technology, magic, and the cultural contexts of technical communication. *Journal of Business and Technical Communication, 14*(3), 289–314.

Lave, J., & Wenger, E. (1991). *Situated learning: Legitimate peripheral participation*. Cambridge, UK and New York: Cambridge University Press.

Mao, L. (2003). Reflective encounters: Illustrating comparative rhetoric. *Style, 37*, 401–425.

Mao, L. (2005). Rhetorical borderlands: Chinese American rhetoric in the making. *College Composition and Communication, 56*(3), 426–469.

Mirel, B., & Spilka, R. (2002). *Reshaping technical communication: New directions and challenges for the 21st century*. London: Routledge.

Murray, H. (2007). Monstrous play in negative spaces: Illegible bodies and the cultural construction of biometric technology. *The Communication Review, 10*, 347–365.

Pennycook, A. (2010). *Language as a local practice*. Milton Park, Abingdon and New York: Routledge.

Rude, C. D. (2004). Toward an expanded concept of rhetorical delivery: The uses of reports in public policy debates. *Technical Communication Quarterly, 13*(3), 271–288. https://doi.org/10.1207/s15427625tcq1303_3.

Savage, G. (2004). Tricksters, fools, and sophists: Technical communication as postmodern rhetoric. In T. Kynell-Hunt & G. Savage (Eds.), *Power and legitimacy in technical communication: Strategies for professional status* (Vol. 2). Amityville, NY: Baywood Publishing Company, Inc.

Scott, B. (2003). *Risky rhetoric: AIDS and the cultural practices of HIV testing*. Carbondale: Southern Illinois University Press.

Simmons, W. M., & Grabill, J. T. (2007). Toward a civic rhetoric for technologically and scientifically complex places: Invention, performance, and participation. *College Composition and Communication, 58*(3), 419–448.

Street, B., & Lefstein, A. (2007). *Literacy: An advanced resource book*. English Language and Applied Linguistics. London: Routledge.

Sullivan, D. L. (2004). Political-ethical implications of defining technical communication as a practice. In J. S. Johnson-Eilola & S. A. Selber (Eds.), *Central works in technical communication* (pp. 211–219). New York: Oxford University Press.

Sun, H. (2009). Toward a rhetoric of locale: Localizing mobile messaging technology into everyday life. *Journal of Technical Writing and Communication, 39*(3), 245–261. https://doi.org/10.2190/TW.39.3.c.

Sun, H. (2012). *Cross-cultural technology design: Creating culture-sensitive technology for local users*. New York: Oxford University Press.

Thatcher, B. (2006). Intercultural rhetoric, technology transfer, and writing in US–Mexico border maquilas. *Technical Communication Quarterly, 15*(3), 385–405.

Thatcher, B. (2010). Editor introduction: Eight needed developments and eight critical contexts for global inquiry. *Rhetoric, Professional Communication, and Globalization, 1*(1), 1–34.

Thatcher, B. (2011). *Intercultural rhetoric and professional communication: Technological advances and organizational behavior*. Hershey, PA: IGI Global Press.

Tuomi, I. (2005). Beyond user-centric models of product creation. In L. Haddon, L. Fortunati, A. Kant, K.-H. Kommonen, E. Mante, & B. Sapio (Eds.), *Everyday innovators: Researching the role of users in shaping ICT's*. Dordrecht: Springer.

Walton, R. (2013). Civic engagement, information technology, and global contexts. *Connexions: An International Professional Communication Journal, 1*(1), 147–154.

Bibliography

1992 Constitution of the Republic of Ghana. (1992).
Acharya, K. R. (2019). Usability for social justice: Exploring the implementation of localization usability in Global North technology in the context of a Global South's country. *Journal of Technical Writing and Communication, 49*(1), 6–32. https://doi.org/10.1177/0047281617735842.
Agboka, G. Y. (2012). Liberating intercultural technical communication from "large culture" ideologies: Constructing culture discursively. *Journal of Technical Writing and Communication, 42*(2), 159–181.
Agboka, G. Y. (2013). Participatory localization: A social justice approach to navigating unenfranchised/disenfranchised cultural sites. *Technical Communication Quarterly, 22*(1), 28–49. https://doi.org/10.1080/10572252.2013.730966.
Agboka, G. Y. (2014). Decolonial methodologies: Social justice perspectives in intercultural technical communication research. *Journal of Technical Writing and Communication, 44*(3), 297–327.
Agboka, G. Y., & Dorpenyo, I. (2018). Technical communication and election technologies. *Technical Communication, 65*(4), 349–352.
Althusser, L. (1971). Ideology and ideological state apparatuses (notes towards an investigation). In L. Althusser (Ed.), *Lenin and philosophy and other essays*. London: New Left Books.
Appadurai, A. (1996). *Modernity at large: Cultural dimensions of globalization* (Vol. 1). Minneapolis: University of Minnesota Press.
Arhin, A. (2016). *CODEO'S interim statement: Observation of the on-going voter register exhibition exercise*. Retrieved from Accra, Ghana.

Aristotle. (2014). On "Techne" and "Episteme". In R. C. Scharff & V. Dusek (Eds.), *Philosophy of technology: The technological condition: An anthology* (2nd ed., pp. 19–22). West Sussex: Wiley Blackwell and Sons.

Atwill, J. M. (1998). *Rhetoric reclaimed: Aristotle and the liberal arts tradition*. Ithaca: Cornell University Press.

Baneseh, M. A. (2015). *Pink sheet: The story of Ghana's presidential election as told in the Daily Graphic*. Accra: G-Pak Limited.

Banks, A. J. (2006). *Race, rhetoric, and technology: Searching for higher ground*. Mahwah, NJ and Urbana, IL: Lawrence Erlbaum and National Council of Teachers of English.

Barton, B. F., & Barton, M. S. (2004). Ideology and the map: Toward a postmodern visual design practice. In J. S. Johnson-Eilola & S. A. Selber (Eds.), *Central works in technical communication* (pp. 232–252). New York: Oxford University Press.

Barton, D., & Hamilton, M. (1998). *Local literacies: Reading and writing in one community*. London: Psychology Press.

Bazerman, C. (1998). The production of technology and the production of human meaning. *Journal of Business and Technical Communication, 12,* 381–387.

Beer, D. (2014). The biopolitics of biometrics: An interview with Btihaj Ajana. *Theory, Culture and Society, 31,* 329–336.

The Biometric Verification Device User Manual. (2012). In T. E. C. o. Ghana (Ed.). Accra.

Bitzer, L. F. (1992). The rhetorical situation. *Philosophy & Rhetoric, 25,* 1–14.

Blyler, N. (1998). Taking a political turn: The critical perspective and research in professional communication. *Technical Communication Quarterly, 7*(1), 33–52.

Boiarsky, C. (1995). The relationship between cultural and rhetorical conventions: Engaging in international communication. *Technical Communication Quarterly, 4*(3), 245–259.

Bowdon, M. (2004). Technical communication and the role of the public intellectual: A community HIV-prevention case study. *Technical Communication Quarterly, 13*(3), 325–340. https://doi.org/10.1207/s15427625tcq1303_6.

Bowen, L. M., Arko, K., Beatty, J., Delaney, C., Dorpenyo, I., Moeller, L., … Velat, J. (2014). Community engagement in a graduate-level community literacy course. *Community Literacy Journal, 9*(1), 18–38.

Brady, A. (2003). Interrupting gender as usual: Metis goes to work. *Women's Studies, 32*(2), 211–233.

Brady, A. (2004). Rhetorical research: Toward a user-centered approach. *Rhetoric Review, 23*(1), 57–74.

Brady, A., & José, L. (2009). Writing for an international audience in a US technical communication classroom: Developing competences to communicate knowledge across cultures. *Nordic Journal of English Studies, 8*(1), 41–60.

Bryant, A., & Charmaz, K. (2007). *Introduction grounded theory research: Methods and practices* (A. Bryant & K. Charmaz, Eds., Paperback ed.). Los Angeles: Sage.

Charmaz, K. (2014). *Constructing grounded theory* (2nd ed.). London; Thousand Oaks, CA: Sage.

Cheeseman, N. (2015). *Democracy in Africa: Successes, failures, and the struggle for political reform* (Vol. 9). Cambridge: Cambridge University Press.

Colman, J. (2015). *Wicked ambiguity and user experience*. Retrieved from http://www.jonathoncolman.org/2015/05/21/wicked-ambiguity/#video.

Colton, J. S., & Holmes, S. (2018). A social justice theory of active equality for technical communication. *Journal of Technical Writing and Communication, 48*(1), 4–30.

Comaroff, J., & Comaroff, J. (1991). Of revelation and revolution. In *Christianity, colonialism, and consciousness in South Africa* (Vol. 1). Chicago: University of Chicago Press.

Coppola, N. W. (2006). Guest editor's introduction: Communication in technology transfer and diffusion: Defining the field. *Technical Communication Quarterly, 15*(3), 285–292.

Cushman, E., & Grabill, J. T. (2009). Writing theories/changing communities: Introduction. *Reflections: Writing, Service-Learning, and Community Literacy, 8*(3), 1–20.

Dean, M. (2010). *Governmentality: Power and rule in modern society*. London: Sage.

Detienne, M., & Vernant, J. P. (1991). *Cunning intelligence in Greek culture and society*. Chicago: University of Chicago Press.

Ding, H. (2014). *Rhetoric of a global epidemic: Transcultural communication about SARS*. Carbondale, IL: Southern Illinois University Press.

Ding, H., & Savage, G. (2013). Guest editors' introduction: New directions in intercultural professional communication. *Technical Communication Quarterly, 22*(1), 1–9.

Dobrin, D. (2004). What's technical about technical writing. In J. S. Johnson-Eilola & S. A. Selber (Eds.), *Central works in technical communication* (pp. 107–123). New York: Oxford University Press.

Doheny-Farina, S. (1992). *Rhetoric, innovation, technology: Case studies of technical communication in technology transfers*. London: MIT Press.

Dolmage, J. (2006). "Breathe upon us an even flame": Hephaestus, history, and the body of rhetoric. *Rhetoric Review, 25*(2), 119–140.

Dolmage, J. (2009). Metis, Metis, Mestiza, Medusa: Rhetorical bodies across rhetorical traditions. *Rhetoric Review, 28*(1), 1–28.

Dong, Q. (2007). *Cross-cultural considerations in instructional documentation: Contrasting Chinese and US home heater manuals*. Paper presented at the

Proceedings of the 25th Annual ACM International Conference on Design of Communication.
Dorpenyo, I. K. (2016). *"Unblackboxing" technology through the rhetoric of technical communication: Biometric technology and Ghana's 2012 election*. Open Access Dissertation, Michigan Technological University.
Dorpenyo, I., & Agboka, G. (2018). Technical communication and election technologies. *Technical Communication, 65*(4), 349–352.
Douglas, T. (2017). To design better tech, understand context. *TEDGlobal 2017*. Retrieved from https://www.ted.com/talks/tania_douglas_to_design_better_tech_understand_context/transcript?language=en#t-503002.
Dourish, P. (2003). The appropriation of interactive technologies: Some lessons from placeless documents. *Computer Supported Cooperation Work, 12*, 465–490.
Dubinsky, J. M. (2004). Guest editor's introduction. *Technical Communication Quarterly, 13*(3), 245–249.
Durack, K. T. (1997). Gender, technology, and the history of technical communication. *Technical Communication Quarterly, 6*(3), 249–260.
Durao, R. (2013). International professional communication: An overview. *Connexions: International Professional Communication Journal, 1*(1), 1–24.
Election Integrity. (2017). *International foundation for electoral systems*. Retrieved August 26, 2017, from http://www.ifes.org/issues/electoral-integrity.
Emerson, R. M., Fretz, R. I., & Shaw, L. L. (2011). *Writing ethnographic fieldnotes* (2nd ed.). Chicago: The University of Chicago Press.
Esselink, B. (2000). *A practical guide to localization* (Vol. 4). Amsterdam: John Benjamins Publishing Company.
Fahnestock, J. (1986). Accommodating science: The rhetorical life of scientific facts. *Written Communication, 3*(3), 275–296.
Final Report on Ghana's 2012 Presidential and Parliamentary Elections. (2013). Retrieved from Accra, Ghana. http://www.codeoghana.org/assets/downloadables/Final%20Report%20on%20Ghana's%202012%20Presidential%20and%20Parliamentary%20Elections.pdf.
Flower, L. (2008). *Community literacy and the rhetoric of public engagement*. Carbondale, IL: SIU Press.
Foucault, M. (2009). *Security, territory, population: Lectures at the Collège de France 1977–1978* (M. Senellart, Ed., Vol. 4). New York: Macmillan.
Foucault, M., Senellart, M., Ewald, F., Fontana, A., & Burchell, G. (2010). *The birth of biopolitics: Lectures at the Collège de France, 1978–1979*. New York: Palgrave Macmillan.
Fukuoka, W., Kojima, Y., & Spyridakis, J. H. (1999). Illustrations in user manuals: Preference and effectiveness with Japanese and American readers. *Technical Communication, 46*(2), 167–176.

Genkey Elections: One Person, One Vote.
Gonzales, L. (2018). *Sites of translation: What multilinguals can teach us about digital writing and rhetoric.* Ann Arbor: University of Michigan Press.
Gonzales, L., & Zantjer, R. (2015). Translation as a user-localization practice. *Technical Communication, 62*(4), 271–284.
Goodall, H. L. J. (2000). *Writing the new ethnography.* New York: Rowman and Littlefield.
Gould, S. J. (1996). *The mismeasure of man.* New York: W. W. Norton.
Grabill, J. T. (2001). *Community literacy programs and the politics of change.* Albany: State University of New York Press.
Grabill, J. T., & Simmons, W. M. (1998). Toward a critical rhetoric of risk communication: Producing citizens and the role of technical communicators. *Technical Communication Quarterly, 7*(4), 415–441. https://doi.org/10.1080/10572259809364640.
Group, J. (Producer). (2014, 10/11). Nana Akufo-Addo's concession speech on the Supreme Court ruling on the 2012 presidential petition.
Gurak, L. J., & Lannon, J. M. (2012). *Strategies for technical communication in the workplace.* New York, NY: Pearson Higher Ed.
Gyimah-Boadi, E. (2004). *Ensuring violence-free December 2004 elections in Ghana: Early warning facilities.* Paper presented at the Briefing Paper, Ghana Center for Democratic Development.
Gyimah-Boadi, E. (2013). *Strengthening democratic governance in Ghana: Proposals for intervention and reform.* Accra: A Publication of Star Ghana.
Haas, A. M. (2012). Race, rhetoric, and technology: A case study of decolonial technical communication theory, methodology, and pedagogy. *Journal of Business and Technical Communication, 26*(3), 277–310.
Haraway, D. (1988). Situated knowledges: The science question in feminism and the privilege of partial perspective. *Feminist Studies, 14*(3), 575–599.
Haraway, D. (1991). *Simians, cyborgs, and women: The reinvention of women.* London and New York: Routledge.
Harding, S. G. (1991). *Whose science? Whose knowledge? Thinking from women's lives.* Ithaca, NY: Cornell University Press.
Hart, R. P., & Daughton, S. M. (2005). *Modern rhetorical criticism* (3rd ed.). New York, USA: Pearson Education.
Heidegger, M. (2014). The question concerning technology. In R. C. Scharff & V. Dusek (Eds.), *Philosophy of technology: The technological condition: An anthology* (2nd ed., pp. 305–317). West Sussex: Wiley Blackwell and Sons.
Hobbis, S. K., & Hobbis, G. (2017). *Voter integrity, trust and the promise of digital technologies: Biometric voter registration in Solomon Islands.* Paper presented at the Anthropological Forum.
Hoft, N. L. (1995). *International technical communication: How to export information about high technology.* New York, NY: Wiley.

Hogan, D. B. (2013). Theories that apply to technical communication. *Connexions: International Professional Communication Journal*, 1(1), 155–165.

Hristova, S. (2014). Recognizing friend and foe: Biometrics, veridiction, and the Iraq War. *Surveillance & Society*, 12(4), 516.

Hunsinger, R. P. (2006). Culture and cultural identity in intercultural technical communication. *Technical Communication Quarterly*, 15(1), 31–48.

Hunsinger, R. P., Amant, K. St., & Sapienza, F. (2011). Using global contexts to localize online contents for international audiences. In *Culture, communication and cyberspace: Rethinking technical communication for international online environments* (pp. 13–37). Amityville, NY: Baywood.

Jockers, H., Kohnert, D., & Nugent, P. (2010). The successful ghana election of 2008: A convenient myth? *The Journal of Modern African Studies*, 48(01), 95–115.

Johnson, R. (1998). *User centered technology: A rhetorical theory for computers and other mundane artifacts*. Albany: State University of New York.

Jones, N. N. (2014). Methods and meanings: Reflections on reflexivity and flexibility in an intercultural ethnographic study of an activist organization. *Rhetoric, Professional Communication, and Globalization*, 5(1), 14–43.

Jones, N. N. (2016a). Narrative inquiry in human-centered design: Examining silence and voice to promote social justice in design scenarios. *Journal of Technical Writing and Communication*, 46(4), 471–492. https://doi.org/10.1177/0047281616653489.

Jones, N. N. (2016b). The technical communicator as advocate: Integrating a social justice approach in technical communication. *Journal of Technical Writing and Communication*, 46(3), 342–361. https://doi.org/10.1177/0047281616639472.

José, L. (2017). *User-centered design and normative practices: The Brexit vote as a technical communication failure*. Paper presented at the Association of Teachers of Technical Writing–Twentieth Annual Conference, Portland, OR.

Katz, S. B. (1992). The ethic of expediency: Classical rhetoric, technology, and the Holocaust. *College English*, 54(3), 255–275.

Keller, E. F. (1996). Feminism and science. In E. F. Keller & H. E. Longino (Eds.), *Feminism and science* (pp. 28–39). New York: Oxford University Press.

Kelly, B., & Bening, R. (2013). The Ghanaian elections of 2012. *Review of African Political Economy*, 40(137), 475–484.

Kimball, M. A. (2006). Cars, culture, and tactical technical communication. *Technical Communication Quarterly*, 15(1), 67–86.

Kimball, M. A. (2017). Tactical technical communication. *Technical Communication Quarterly*, 26(1), 1–7. https://doi.org/10.1080/10572252.2017.1259428.

Kitalong, K. S. (2000). "You will": Technology, magic, and the cultural contexts of technical communication. *Journal of Business and Technical Communication, 14*(3), 289–314.

Koerber, A. (2000). Toward a feminist rhetoric of technology. *Journal of Business and Technical Communication, 14*(1), 58–73.

Latour, B. (1987). *Science in action: How to follow scientists and engineers through society*. Cambridge, MA: Harvard University Press.

Lave, J., & Wenger, E. (1991). *Situated learning: Legitimate peripheral participation*. Cambridge, UK and New York: Cambridge University Press.

Light, A., & Luckin, R. (2008). *Designing for social justice: People, technology, learning*. Retrieved from https://www.nfer.ac.uk/publications/FUTL53/FUTL53.pdf.

Lindlof, T. R., & Taylor, B. C. (2011). *Qualitative communication research methods* (3rd ed.). Los Angeles: Sage.

Longo, B. (2000). *Spurious coin: A history of science, management, and technical writing*. Albany: State University of New York Press.

Longo, B., & Fountain, K. (2013). What can history teach us about technical communication? In J. Johnson-Eilola & S. A. Selber (Eds.), *Solving problems in technical communication* (pp. 165–186). Chicago: The University of Chicago Press.

Lovitt, C. R., & Goswami, D. (1999). *Exploring the rhetoric of international professional communication: An agenda for teachers and researchers*. Amityville, NY: Baywood Publishing Company.

Magnet, S. (2011). *When biometrics fail: Gender, race, and the technology of identity*. Durham, NC: Duke University Press.

Maguire, M. (2009). The birth of biometric security. *Anthropology Today, 25*(2), 9–14.

Makulilo, A. (2017). Rebooting democracy? Political data mining and biometric voter registration in Africa. *Information & Communications Technology Law, 26*(2), 198–212.

Manuh, T. (2011). Towards greater representation of women in national governance. *Governance Newsletter*, 17. Accra.

Mao, L. (2003). Reflective encounters: Illustrating comparative rhetoric. *Style, 37*, 401–425.

Mao, L. (2005). Rhetorical borderlands: Chinese American rhetoric in the making. *College Composition and Communication, 56*(3), 426–469.

Markel, M. (2013). *Practical strategies for technical communication*. Boston: Bedford/St. Martin.

McKeon, R. (2009). *The basic works of Aristotle*. New York: Random House LLC.

Miller, C. R. (1978). Technology as a form of consciousness: A study of contemporary ethos. *Communication Studies, 29*(4), 228–236.

Miller, C. R. (1994). Opportunity, opportunism, and progress: Kairos in the rhetoric of technology. *Argumentation, 8*(1), 81–96.
Miller, C. R. (1998). Learning from history: World War II and the culture of high technology. *Journal of Business and Technical Communication, 12*(3), 288–315.
Mirel, B., & Spilka, R. (2002). *Reshaping technical communication: New directions and challenges for the 21st century*. London: Routledge.
Mision, Vision and Functions.
Murray, H. (2007). Monstrous play in negative spaces: Illegible bodies and the cultural construction of biometric technology. *The Communication Review, 10,* 347–365.
MyJoyOnline. (2013). Election 2012 petition hearing-day 2. *Election 2012 petition hearing.*
MyJoyOnline (Producer). (2015, November 21). NPP affirmative action scrapped; November Congress to Consider Adoption.
Norman, D. A. (1988). *The psychology of everyday things*. New York: Basic Books.
Norman, D. (2013). *The design of everyday things: Revised and expanded edition.* New York: Basic Books.
Norman, D., & Nielsen, J. (2008). *The definition of user experience.* Retrieved from https://www.nngroup.com/articles/definition-user-experience/.
Norris, P. (2014). *Why electoral integrity matters.* New York: Cambridge University Press.
Oliu, W. E. (1994). *Writing that works: Effective communication in business.* Scarborough, ON: Nelson.
Oquaye, M. (1995). The ghanaian elections of 1992—A dissenting view. *African Affairs, 94*(375), 259–275.
Oquaye, M. (2004). *Politics in Ghana, 1982–1992: Rawlings, revolution, and populist democracy.* Accra and New Delhi: Tornado Publications.
Oquaye, M. (2012a). *Reserving special seats for women in parliament: Issues and obstacles.* Accra: Governance Newsletter.
Oquaye, M. (2012b). *Strengthening Ghana's electoral system: A precondition for stability and development.* IEA Monograph (No. 38). Ghana.
Paradis, J. (2004). Text and action: The operator's manual in context and in court. In J. S. Johnson-Eilola & S. A. Selber (Eds.), *Central works in technical communication* (pp. 365–380). New York: Oxford University Press.
Parliamentary Debates: Official Report (Emergency Meeting). (2012). Retrieved from Parliament House, Accra.
Pennycook, A. (2010). *Language as a local practice.* Milton Park, Abingdon and New York: Routledge.
Pfeiffer, W. S. (2003). *Technical writing: A practical approach.* Toronto: Pearson College Division.

Preliminary Report on Monitoring the Right to Vote and Observing the 2012 Presidential and Parliamentary Elections in Ghana. (2012). Retrieved from Accra.

Pugliese, J. (2010). *Biometrics: Bodies, technologies, biopolitics* (Vol. 12). New York: Routledge.

Reed, T. V. (2014). *Digitized lives: Culture, power, and social change in the Internet era.* New York: Routledge.

Report of the Commonwealth Observer Group: Ghana Presidential and Parliamentary Elections. (2013). Retrieved from London. http://thecommonwealth.org/sites/default/files/inline/GhanaElections-FinalReport2012.pdf.

Rose, E. J. (2016). Design as advocacy: Using a human-centered approach to investigate the needs of vulnerable populations. *Journal of Technical Writing and Communication, 46*(4), 427–445. https://doi.org/10.1177/0047281616653494.

Roy, D. (2013). Toward experience design: The changing face of technical communication. *Connexions: International Professional Communication Journal, 1*(1), 111–118.

Rubin, H. J., & Rubin, I. S. (2005). *Qualitative interviewing: The art of hearing data* (2nd ed.). Thousand Oaks, CA: Sage.

Rude, C. D. (2004). Toward an expanded concept of rhetorical delivery: The uses of reports in public policy debates. *Technical Communication Quarterly, 13*(3), 271–288. https://doi.org/10.1207/s15427625tcq1303_3.

Rude, C. D. (2009). Mapping the research questions in technical communication. *Journal of Business and Technical Communication, 23*(2), 174–215.

Rutter, R. (1991). History, rhetoric, and humanism: Toward a more comprehensive definition of technical communication. *Journal of Technical Writing and Communication, 21*(2), 133–153.

Saaka, Y. (1997). Legitimizing the illegitimate: The 1992 presidential elections as a prelude to Ghana's fourth republic. *Issues and trends in contemporary African politics* (pp. 143–172). New York: Peter Lang.

Saldana, J. (2009). *The coding manual for qualitative researchers.* Thousand Oaks, CA: Sage.

Salia, A. (2000). Voters Register in excess of 1.5 m people—Afare-Gyan. *Daily Graphic, 21*, 2000.

Salvo, M. J. (2001). Ethics of engagement: User-centered design and rhetorical methodology. *Technical Communication Quarterly, 10*(3), 273–290.

Sapp, D. A., Savage, G., & Mattson, K. (2013). After the International Bill of Human Rights (IBHR): Introduction to special issue on human rights and professional communication. *Rhetoric, Professional Communication, and Globalization, 14*(1), 1–12.

Savage, G. (2004). Tricksters, fools, and sophists: Technical communication as postmodern rhetoric. In T. Kynell-Hunt & G. Savage (Eds.), *Power and legitimacy in technical communication: Strategies for professional status* (Vol. 2). Amityville, NY: Baywood Publishing Company, Inc.

Scott, B. (2003). *Risky rhetoric: AIDS and the cultural practices of HIV testing*. Carbondale: Southern Illinois University Press.

Seigel, M. (2006). *Reproductive technologies: Pregnancy manuals as technical communication*. Ann Arbor: ProQuest.

Seigel, M. (2013). *The rhetoric of pregnancy*. Chicago: University of Chicago Press.

Selfe, C. L. (1999). *Technology and literacy in the twenty-first century: The importance of paying attention*. Carbondale, IL: SIU Press.

Simmons, W. M., & Grabill, J. T. (2007). Toward a civic rhetoric for technologically and scientifically complex places: Invention, performance, and participation. *College Composition and Communication, 58*(3), 419–448.

Slack, J. D., & Wise, J. M. (2005). *Culture and technology: A primer*. New York: Peter Lang.

Smith, B. (2012). *Reading and writing in the global workplace: Gender, literacy, and outsourcing in Ghana*. Lanham: Lexington Books.

Smith, L. T. (2012). *Decolonial methodology: Research and indigenous peoples* (2nd ed.). London: Zed Books.

Spinuzzi, C. (2003). *Tracing genres through organizations: A sociocultural approach to information design* (Vol. 1). Cambridge, MA: MIT Press.

Spinuzzi, C. (2013). *Topsight: A guide to studying, diagnosing, and fixing information flow in organizations*. CreateSpace Independent Publishing Platform.

Spinuzzi, C., & Zachry, M. (2000). Genre ecologies: An open-system approach to understanding and constructing documentation. *ACM Journal of Computer Documentation (JCD), 24*(3), 169–181.

St. Amant, K. (2017). Of scripts and prototypes: A two-part approach to user experience design for international contexts. *Technical Communication, 64*(2), 113–125.

stlghana. (2012). *STL Ghana—Biometric system voter registration process*.

Strauss, A. L., & Corbin, J. M. (1990). *Basics of qualitative research: Grounded theory procedures and techniques*. Newbury Park, CA: Sage.

Street, B., & Lefstein, A. (2007). *Literacy: An advanced resource book*. English Language and Applied Linguistics. London: Routledge.

Sullivan, D. L. (2004). Political-ethical implications of defining technical communication as a practice. In J. S. Johnson-Eilola & S. A. Selber (Eds.), *Central works in technical communication* (pp. 211–219). New York, NY: Oxford University Press.

Sullivan, P., & Porter, J. E. (1997). *Opening spaces: Writing technologies and critical research practices*. Greenwich: Greenwood Publishing Group.

Sun, H. (2004). *Expanding the scope of localization: A cultural usability perspective on mobile text messaging use in American and Chinese contexts*. Troy, NY: Rensselaer Polytechnic Institute.

Sun, H. (2006). The triumph of users: Achieving cultural usability goals with user localization. *Technical Communication Quarterly, 15*(4), 457–481. https://doi.org/10.1207/s15427625tcq1504_3.

Sun, H. (2009a). Designing for a dialogic view of interpretation in cross-cultural IT design. In N. Aykin (Ed.), *Internationalization, design and global development* (pp. 108–116). Heidelberg: Springer.

Sun, H. (2009b). Toward a rhetoric of locale: Localizing mobile messaging technology into everyday life. *Journal of Technical Writing and Communication, 39*(3), 245–261. https://doi.org/10.2190/TW.39.3.c.

Sun, H. (2012). *Cross-cultural technology design: Creating culture-sensitive technology for local users.* New York: Oxford University Press.

Supreme Court Judgement on the Presidential Election Petition, Akufo-Addo V. Mahama and Others (A.D 2013). Retrieved from http://www.myjoyonline.com/docs/full-sc-judgement.pdf.

Taylor, D. (1992). *Global software: Developing applications for the international market.* New York: Springer.

Thatcher, B. (2006). Intercultural rhetoric, technology transfer, and writing in US–Mexico border maquilas. *Technical Communication Quarterly, 15*(3), 385–405.

Thatcher, B. (2010). Editor introduction: Eight needed developments and eight critical contexts for global inquiry. *Rhetoric, Professional Communication, and Globalization, 1*(1), 1–34.

Thatcher, B. (2011). *Intercultural rhetoric and professional communication: Technological advances and organizational behavior.* Hershey, PA: IGI Global.

Thayer, A., & Kolko, B. E. (2004). Localization of digital games: The process of blending for the global games market. *Technical Communication, 51*(4), 477–488.

Thrush, E. A. (1993). Bridging the gaps: Technical communication in an international and multicultural society. *Technical Communication Quarterly, 2*(3), 271–283.

Tracy, S. J. (2010). Qualitative quality: Eight "big-tent" criteria for excellent qualitative research. *Qualitative Inquiry, 16*(10), 837–851.

Tuomi, I. (2005). Beyond user-centric models of product creation. In L. Haddon, L. Fortunati, A. Kant, K.-H. Kommonen, E. Mante, & B. Sapio (Eds.), *Everyday innovators: Researching the role of users in shaping ICT's.* Dordrecht: Springer.

User Experience Network. (2008). Retrieved from http://www.uxnet.org/.

Utley, I. (2009). *The essential guide to customs and culture: Ghana.* London: Kuperard.

Vacca, J. R. (2007). *Biometric technologies and verification systems.* Burlington, MA: Butterworth-Heinemann.

van Reijswoud, V., & de Jager, A. (2011). The role of appropriate ICT in bridging the digital divide. In K. St. Amant & B. A. Olaniran (Eds.), *Globalization and the digital divide*. Amherst: Cambria Press.

Wajcman, J. (1991). *Feminism confronts technology*. Pennsylvania: The Pennsylvania State University Press.

Walton, R. (2013a). Civic engagement, information technology, and global contexts. *Connexions: An International Professional Communication Journal*, *1*(1), 147–154.

Walton, R. (2013b). How trust and credibility affect technology-based development projects. *Technical Communication Quarterly*, *22*(1), 85–102.

Walton, R. (2014). Editor's introduction to the special edition on methodology. *Rhetoric, Professional Communication, and Globalization*, *5*(1), 1–13.

Walton, R. (2016). Supporting human dignity and human rights: A call to adopt the first principle of human-centered design. *Journal of Technical Writing and Communication*, *46*(4), 402–426. https://doi.org/10.1177/0047281616653496.

We Believe That Democracy Is a Design. (2019). Retrieved from http://civicdesign.org/about/.

Wells, S. (2010). Technology, genre, and gender. In S. A. Selber (Ed.), *Rhetorics and technologies: New directions in writing and communication*. Carolina, SC: The University of South Carolina Press.

Whitney, J. G. (2013). The 2010 citizens clean elections voter education guide: Constructing the "Illegal Immigrant" in the Arizona voter. *Journal of Technical Writing and Communication*, *43*(4), 437–455.

Winner, L. (1980). Do artifacts have politics? *Daedalus*, *109*(1), 121–136.

Winner, L. (1986). *The whale and the reactor: A search for limits in an age of high technology (La Baleine et le réacteur)*. Chicago: Chicago University Press.

Young, I. M. (1990). *Justice and the politics of difference*. Princeton, NJ: Princeton University Press.

Index

A
Access, 49, 59, 61, 62, 67, 88, 89, 123, 140, 186, 189, 190, 193, 195, 206
Accra, 37, 40, 41, 62
Accuracy, 17, 18, 31, 48–50, 64, 94, 106, 107, 112–116, 120, 121, 125, 159, 190, 191
Actions of users, 143
Active participation, 10, 202, 204
Adam Banks, 189
Adopt and use, 12, 20, 96, 159, 166
Adopting, 3, 7, 42, 47, 50, 71, 89, 137, 192, 206
Adoption and use of technology, 3, 11, 22, 211
Afari-Gyan, Kwadwo (Dr), 44, 47, 147, 151, 178
Africa, 14, 28, 29, 38, 42–47, 107
African countries, 14, 47
African Union, 43
African Union Observation Report, 15
Agboka, Godwin, 5, 6, 10, 12, 17, 22–25, 40, 54, 56, 58, 63, 91, 92, 103, 165, 167, 173, 188, 192, 203–205
Agency, 6, 8, 12, 23, 110, 133, 143, 144, 164, 192, 209
Agency of users, 7
Akan, 41, 116, 117
Akufo-Addo, Nana, 47
Angela Haas, 189
Annan, Kofi, 43
Appadurai, Arjun, 105, 106
Aristotle, 117, 118, 146, 169, 170
Articulation, 71, 75, 119, 203, 206, 210
Asante, 41
Authentication, 2, 49, 118, 207

B
Ballot box snatching, 2
Bazerman, Charles, 18, 103, 104, 116, 149
Beacon of democracy, 42
Behavioral, 49, 50, 114, 118, 216
Biometric, 2, 7, 8, 10, 13, 14, 16–21, 27–31, 38, 40, 46–50, 63, 65,

68, 69, 71, 75, 79, 81, 83–90,
93–100, 103–126, 129–132, 134,
136–140, 142, 143, 145–147,
149–154, 156, 158, 159, 161,
163, 164, 166, 168, 175, 176,
186, 188–191, 193–198, 202,
204–207, 210, 212, 213, 215,
216
Biometric as a tool of disenfranchisement, 187
Biometric breakdowns, 3, 9, 10, 14,
19, 26, 31, 74, 82, 84, 134, 135,
138, 143, 147, 194
Biometric election, 16, 21, 23, 32,
46, 50, 88, 89, 92, 93, 109, 113,
119, 121, 125, 143, 146, 168,
186, 193, 196, 216, 217
Biometric ethos, 112, 117, 129
Biometric fails to recognize your fingers, 2
Biometric failure, 10, 25, 30, 31, 129,
130, 138, 143, 194
Biometric ideology, 17, 30, 106, 108,
109, 121, 125, 190
Biometric law, 95, 112, 121, 207
Biometric rejection, 26, 188, 193, 196
Biometric technology, 2, 3, 7–9, 14,
16, 19, 20, 23, 24, 26–32, 38,
47, 49, 50, 53, 63, 65, 69–71,
73, 75, 79, 80, 82, 83, 86–90,
92, 93, 95, 96, 98, 100, 103–
105, 108, 109, 111, 113–115,
117, 118, 120, 121, 123–126,
129–132, 136–143, 145, 158,
163, 164, 168, 172, 182,
185–190, 194, 196, 198, 202,
207, 208, 211, 213, 215–217
Biometric use, 3, 8, 15, 17, 19, 21,
22, 26, 28, 47, 69, 81, 83, 89,
96, 105, 109, 131, 132, 147,
151, 168, 182, 186, 193, 196,
202, 204, 207

Biometric verification device (BVD),
2, 3, 16, 82, 84, 86, 101, 109,
111, 130, 132, 141, 151, 152,
158, 159, 215
Biometric verification registration
(BVR), 16, 130
Black box, 197
Bodily deviance, 193
Bottom-up, 69, 92, 145
Broader context, 3
Bunch of dummies, 17
Burkina Faso, 41

C
Cable News Network (CNN), 37
Cast a ballot, 23
Cheeseman, Nic, 44–46
Chieftaincy, 59, 116, 117
Civic engagement, 21, 23, 24, 187,
213–215
Civil society organizations, 18
Clash of cultures, 4
Climate control, 14
Clinton, Bill, 42
Coalition of Domestic Election
Observers (CODEO), 130, 141
Cocoa, 32, 42, 85
Collaborative effort, 9, 12
Collectivist, 59, 174, 175, 205
Colman, Jonathan, 4
Colonialism, 30, 55–57, 59, 60, 63,
196
Colonized context, 29, 30, 55–57, 72,
202
Colonizing users, 92
Colony of Britain, 42
Commission on Human Rights and
Administrative Justice (CHRAJ),
26, 32, 134, 186
Complex context of use, 11, 37, 53
Complex nature of use, 16

Constitutional Instrument, 95, 96, 112, 122
Constructed meaning, 15, 131
Constructing local/global knowledge, 91
Consumer site, 166
Context, 3, 7, 9, 10, 13, 15, 16, 21–24, 26, 29–32, 38, 54–60, 64, 66, 73, 89, 90, 100, 103, 110, 118, 119, 121, 125, 131, 132, 135, 137, 139, 140, 145, 146, 148, 149, 163–165, 167, 173, 178, 179, 182, 188, 189, 191–194, 201–209, 211, 214–217
Context of use, 5, 37
Control temperature, 134
Conventions People Party (CPP), 43
Corrupt, 39, 70, 83
Cote D'Ivoire, 41, 42, 45
Coup d'état, 44, 45
Creating technology for users, 9
Creative activities, 20
Creative efforts of users, 7
Critical, 53, 54, 60, 61, 63, 67, 96, 123, 139, 140, 142, 143, 148, 150, 158, 197, 202, 208
Cross-cultural audiences, 12, 23
Cross-cultural context, 12, 24, 29, 54, 165, 188
Cross-cultural design, 9, 15, 24, 131
Cultural expectation of users, 18
Cultural imperialism, 59
Cultural sites, 23, 166
Culture, 4–7, 9, 11, 23, 27, 28, 41, 54–60, 68, 72, 75, 91, 92, 103, 106, 115, 117, 118, 123, 126, 132, 133, 167, 173–175, 178, 179, 182, 187–189, 191, 194, 201–203, 209, 210, 215
Culture in abstract terms, 6
Culture of design, 4, 5

Cunning intelligence, 146, 170, 172, 207
Cunning knowledge, 137
Customization, 4, 5, 19, 136, 143, 189
Cyborg technology, 118–120

D
Dave, Taylor, 4, 5, 16, 61
Decolonial methodology, 28, 30, 53, 56, 57, 59–61, 67, 71, 72, 209
Decolonized, 60
Decontextualized, 132, 205
Democracy, 22, 26, 42, 44, 45, 59, 105–107, 117, 124–126, 211
Demographic failures, 186, 196
Design challenges, 9, 145
Designer as expert of technology, 11
Designers, 3–6, 9–16, 19–21, 26, 30, 50, 59, 60, 85, 92, 100, 103, 115, 131, 133, 136, 139, 142, 143, 145, 146, 148, 150, 164, 167, 168, 170, 188, 189, 191–194, 202, 203, 205, 208, 216
Designing for global use, 3
Deterministic discourses, 17–19, 198
Developer localization, 4, 7, 191
Developing democracies, 38
Diamond, 42
Digital technology, 38, 47, 189
Discrimination inherent in technology design, 188, 189
Disenfranchised, 2, 58, 59, 83, 129, 193, 197
Documentation, 5, 26, 92, 108, 111, 140, 148, 149, 151, 159, 164, 192
Documents, 6, 18, 24, 25, 27, 28, 30, 43, 62, 66–69, 71, 75, 92–97, 100, 103, 104, 108–111, 113, 125, 143, 146–149, 151, 152,

163–168, 170, 172–175, 182, 209, 211
Dominant culture, 54, 56
Dorpenyo, I., 23, 25, 61, 62, 68, 197
Douglas, Tania, 3, 14
Dourish, P., 8, 216, 217
Dummies, 145, 192, 216
Dusty, 2, 15, 41, 85, 88
Dynamic relationship between rhetoric and culture, 53

E

Effectiveness, 50, 141
Efficient, 18, 23, 50, 89, 104, 107–111, 126, 129, 151, 152
Election petition, 47, 82
2012 elections, 7, 47, 65, 69, 74, 82, 100, 114, 136, 140, 141, 143, 147, 204, 213
2016 elections, 10, 69, 70, 74, 84, 141, 204
Elections, 1, 2, 8, 14, 16, 17, 20–26, 31, 38–40, 42, 44–48, 62, 65, 66, 68, 69, 73, 80–87, 89, 90, 95, 96, 106, 109, 112–115, 120, 124, 125, 129, 130, 141, 147, 158, 159, 163, 166, 176, 182, 185, 187, 189, 202, 204, 207, 209, 212, 213, 215
Election technology, 8, 21, 23, 25, 204
Electoral Commission of Ghana (EC), 2, 3, 8–10, 12–14, 19, 21, 28–31, 38–40, 43, 44, 47, 48, 64, 66, 69, 70, 73, 74, 79–87, 92, 93, 95, 104, 106, 108, 110, 118, 126, 131, 133, 134, 137, 139–141, 143, 145–147, 164, 168, 172, 194, 204, 207, 215, 217
Electoral fraud, 38
Electoral issues, 22, 25
Electoral malpractices, 2, 39, 42–45, 47, 89, 96, 105, 107, 108, 115
Electoral process, 2, 19, 22, 28, 31, 37–40, 43, 47, 48, 50, 62, 74, 80–85, 87, 89, 92, 93, 96, 110–112, 115, 125, 129, 130, 132, 140, 151, 158, 168, 182, 193, 194, 202, 204, 205, 209, 213
Electoral system, 9, 10, 16, 19–21, 25, 26, 38–40, 43, 47, 50, 80, 81, 89, 103, 106–109, 112, 126, 130, 136, 187, 202, 217
Electoral woes, 7, 39, 48, 83, 104, 105, 112
Empower, 23, 58, 59, 68, 72, 185, 193
Empowerment, 30, 57, 75
Environmental factors, 11, 14, 203
Esselink, B., 4, 5, 16, 91
Ethic of expediency, 18, 107, 118, 190, 194, 195
Ethics, 22, 54, 169
Ethics of use, 194
Ethos, 47, 110, 112, 115, 117, 118, 129, 195, 196
Ewe, 41
Exclusion, 197
Expertise of local users, 10
Expert knowledge, 135
Expose electoral offenders, 2, 7, 46

F

Fail-safe, 129
Finger print biometric, 118, 151
Finger prints, 3, 26, 48, 85, 97, 110–113, 119, 120, 122, 132, 147, 151, 152, 158, 182, 185, 186, 193–195
Fingers, 2, 26, 48, 49, 73, 87, 122, 134, 159, 175, 186, 191, 193

Fit, 5, 9, 10, 15, 17–20, 24, 75, 84, 103, 113, 136, 143, 145, 146, 149, 168, 189, 193, 201
Fit local context, 17
Foucault, M., 114, 120, 121, 212
Fourth republic, 39, 42
Free and fair elections, 47, 110, 115
Frustrated user, 133, 134

G

Gap between designers and users, 4, 7, 164
Ghana, Kofi, 94, 113, 116–118, 120–124, 210
Ghana, 1, 3, 4, 7–10, 13–19, 21, 23, 26, 28–32, 37–48, 53, 59, 61–64, 66, 69, 75, 79, 80, 82–84, 86, 89, 90, 92, 93, 95–97, 100, 103, 106–110, 112–126, 130, 136, 143, 145, 146, 148, 149, 151, 164, 166, 174, 175, 178, 182, 185, 187, 193–195, 201, 202, 205–207, 209, 210, 212, 213, 215
Global flows, 209
Globalization, 54, 58, 131, 215
Global logics, 29, 38
Global rhetoric, 54, 100, 125
Global users, 3, 4
Gold, 40, 42, 80, 190
Gold Coast, 40
Graphical features, 91, 166
Grounded theory, 30, 71–75

H

Harassment of polling officers, 2
Harmattan, 13, 15, 41, 85
Hazy, 15, 41
Heroes, 17, 29, 138, 145, 208

Heuristic, 11, 17, 18, 31, 39, 53, 57, 71, 135, 137, 138, 142, 202
Heuristic approaches, 17, 19, 31, 55, 133, 134, 146, 206
High context, 179
High temperature, 3, 13, 14, 84, 130
Historical, 22, 29, 38, 40, 44, 92, 93, 123, 124, 158, 191, 209, 210, 216
Hobbis, G., 16, 48, 50, 105, 130, 132, 186
Hobbis, S.K., 16, 48, 50, 105, 130, 132, 186
Hoft, Nancy, 4, 6, 16
Hot weather, 73, 81
Human body, 113, 118, 120, 129, 190, 194
Human centered approach, 191
Human dignity, 21, 187, 194, 195, 208
Human rights, 21, 22, 25, 32, 187, 195, 208
Human thresholds, 175, 210
Humidity, 3, 14, 130
Humid weather, 2
Humility, 65, 66, 72

I

Iceberg model, 6
Identification, 49, 88, 105, 112, 114, 115, 122, 125, 186, 198
Identity, 25, 49, 60, 63, 64, 68, 71, 89, 116, 118, 122, 125, 129, 143, 151, 166, 187, 190, 196, 212
Ideologies embedded in design, 26
Ideology, 17, 30, 58, 64, 105, 106, 110, 122, 191, 197, 212
Illegible biometric bodies, 186
Illiteracy rate, 39, 83

Impersonation, 2, 46, 47, 65, 66, 83, 87, 89, 110, 112, 151
Incontrovertible, 38, 106, 113, 120, 209
Incorporation of technology, 8
Independence, 42–44, 173
Individualistic culture, 174
Inequality, 173, 178, 179, 195
Insider, 63, 64, 67, 210
Institutions, 11, 22, 28, 43, 46, 59, 61, 81, 104, 106, 109, 115, 135, 209, 212
Instructional procedures, 2
Instruction manual, 3, 75, 133, 150, 152, 209
Instructions, 26, 65, 69, 81, 84, 87, 90, 102, 104, 108, 111, 141, 145–152, 158–163, 166, 174, 177, 182
Instrumental affordances, 9
Instrumental features, 17, 18, 30, 89, 103, 126, 148, 204, 217
Instrumentality, 19, 107, 108
Integration of technology, 204
Intended purpose of technology, 8
Intercultural, 23, 29, 54–56, 67, 202
International audiences, 167, 182
International context, 23–25, 53, 55, 56, 201, 211
Internationalization, 4, 5
International technical communication, 24, 29, 54, 58, 63, 64, 66, 173, 211
International Variables Worksheet, 6
Inter Party Advisory Committee (IPAC), 18, 81

J
Johnson, Robert, 10–13, 16, 20–22, 24, 27, 32, 60, 117, 133–135, 149, 163, 168–170, 205, 213
Jones, Natasha, 12, 63, 67, 188, 190, 193, 194, 197
1st of July 1960, 42
Justice, 113, 175, 212

K
Kairos, 111, 170, 171, 207
Kairotic, 48, 96, 112, 115, 207, 214
Katz, Steven, 18, 107, 110, 118, 188, 190, 195
Kimball, Miles, 10, 22, 93, 100, 208, 209
Knowing, 8, 30, 55–57, 60, 72, 85, 92, 134, 137, 169, 170, 172, 192, 211, 212
Knowledge is contingent, 135
Knowledge of use, 7, 12, 13, 22, 88, 192
Knowledge systems, 91, 92
Kufuor, John Agyekum, 42

L
Land of the mundane, 11, 135
Language, 5–8, 16–18, 22, 26, 28, 31, 57, 60, 72, 91, 92, 100, 103, 104, 122, 123, 140, 150, 159, 166, 170, 178, 191, 210, 212, 214
Language use, 8, 92
Large cultural characteristics, 6
Large culture ideology, 54
Learning, 10, 31, 84, 135, 137, 139–142, 163, 205–207, 215
Levels of localization, 4, 10, 164, 166
Linguistic features, 5, 91, 104
Linguistic localization, 10, 18, 28–31, 103, 104, 140, 202
Lived experiences, 28, 68, 194
Local agent, 5
Local culture, 4, 13, 54, 56, 93

Local document, 109, 113
Locale, 5, 6, 92, 105, 146, 216, 217
Local-global tensions, 22
Local herbs, 3, 26, 186
Localization, 4–10, 12, 15–17, 19, 21, 22, 24, 26, 28–32, 53, 58, 60, 61, 70, 72, 75, 88, 89, 91, 92, 100, 103, 109, 130–133, 136, 145, 146, 164–167, 185, 188, 189, 191–193, 195, 197, 202–208, 216
Localization agent, 167
Localization as process, 5, 8–10, 15, 21, 24, 29, 30, 37, 59, 89–91, 132, 138, 139, 144–146, 165–167, 203, 205
Localization as product, 5, 9, 131, 132, 164, 167, 203
Localization cycle, 202, 203
Localization firm, 5
Localization Industry Standards Association (LISA), 5
Localization process, 6
local measures, 3
Local rhetorical cultural values, 5
Local technology use, 8, 24, 38, 100, 103, 211
Local use, 8
Location and place of technology use, 3
Logic of technology, 22, 108, 110
Longo, Bernadette, 23, 27, 103
Lost voices of users, 30, 37, 57, 170
Low context, 179, 182

M
Macho men, 40, 80
Magical, 116, 209
Mahama, John Dramani, 47
6th March 1957, 42
Marginalization, 59, 194, 197, 198

Means to an end, 18, 107, 110, 118
Mental models, 4
Metis, 170–172, 207
Military upheaval, 42
Miller, Carolyn, 18, 92, 93, 103, 107, 111, 116, 216
Minors voting, 9
Model of democracy, 42
Monolithic, 6, 56
Multiparty democracy, 44
Multiple voting, 2, 46, 47, 81, 83, 87, 110, 126, 151
Murray, H., 186, 187, 190, 208

N
Narrow and static definition of culture, 6
Narrow representation of users, 37
National Democratic Congress (NDC), 1, 40, 42, 44, 46
Needs of users, 3, 7, 12, 140
Neocolonialism, 43
Neutral, 17, 18, 26, 47, 50, 106, 107, 110, 122, 188–191, 194, 197
Nkrumah, Kwame (Dr.), 43
No biometric no vote, 113, 120
Non-Western, 21–26, 28, 29, 32, 56, 64, 92, 115, 118, 119, 121, 123, 126, 173, 188, 191, 202, 210, 211
Norman, Don, 104, 130, 133, 134, 138, 139

O
Obama, Barack, 42
Objective, 17, 18, 31, 58, 61, 100, 106, 107, 122, 129, 174, 189–191, 194, 196
OMO, 3, 10, 19, 73, 136, 137

Online video, 28, 30, 68, 69, 75, 93, 96, 100, 103, 104, 108–110, 114–116
Operational affordances, 9
Oppression, 59, 191
Organization of African Unity, 43
Outsider, 63, 64, 68, 197, 210
Over voting, 42, 44, 47, 65, 66, 82, 86, 141

P
Paper ballot, 22, 110, 112, 113, 120, 151, 182
Paradis, James, 27, 147–149, 164
Parliament of Ghana, 18, 46, 95, 104, 113, 120, 207
Participatory citizen, 22, 213
Participatory localization, 17, 24, 165, 203
Participatory user localization, 32, 201, 203
Passive consumers, 6
Patience, 66, 67
Persuasive language, 17
Physical environment, 3, 37
Physiological, 48–50, 118
Pink sheets, 22, 82, 141
Place, 1, 6, 11, 14, 22, 37, 56–59, 63, 68, 86, 88, 94, 100, 106, 121, 135–137, 140, 152, 159, 166, 174, 175, 179, 209, 211, 216, 217
Political, 1, 6, 17, 18, 22, 24–26, 29, 37–40, 42, 43, 45, 46, 53, 55, 60, 80, 81, 83, 86, 92, 106, 108, 110, 113–116, 119–121, 123, 124, 132, 158, 159, 189, 196, 203, 204, 209, 212–216
Political context, 24, 32
Politics, 15, 37, 38, 44, 45, 72, 106, 114, 124, 125, 212, 215

Politics of Ghana, 213
Polling center, 2
Polling station, 2, 3, 10, 13–15, 65, 69, 82, 84, 85, 99, 110, 113, 120, 129, 130, 134, 138, 141, 147, 149, 151, 179, 217
Poor user experience, 6, 19
Porter, James, 53
Positionality, 63, 64
Postcolonial, 37, 38, 40, 64, 66
Post-election conflict, 42, 44
Powerful use of language, 7
Practical knowledge, 118, 172, 207
Practical wisdom, 19, 31, 146, 172, 216
Prevent incidence of multiple voting, 7
Procedures, 19, 31, 62, 65, 67, 73, 82, 99, 104, 135–137, 147, 149, 151, 158, 163, 181, 182
Producer site, 166
Product designers, 4
Product designers versus user of product, 4, 7, 164
Product Users, 4
Professor Evans Atta Mills (late), 42
Provide transparency, 2, 7, 46
Pugliese, J., 49, 50, 114, 118, 120

R
Racial discrimination, 10, 26
Rawlings, Jerry John, 45, 46
Reciprocity, 201
Reconfigure, 6–8, 89, 145, 146, 192, 202
Recovery process, 57
Reflective, 11, 53, 142, 205, 211
Reflexivity, 67, 68, 72
Rejected voter, 10, 73, 196, 197
Research, 6, 12, 20, 22–26, 28–30, 32, 53–55, 57–72, 142, 145,

173, 186, 188, 191, 192, 198, 208, 209, 211, 214, 215
Research methodology, 53, 64
Resource-mismanaged context, 53
Rhetoric, 7, 18, 29, 50, 55, 75, 93, 103, 104, 114, 149, 163, 171, 173, 188, 189, 202, 212, 214
Rhetorical, 5, 7, 11, 12, 16, 18, 22, 23, 25, 28, 38, 45, 53–55, 92, 104, 111, 123, 125, 126, 129, 139, 171, 173–175, 182, 201, 202, 205, 207, 211, 212, 215–217
Rhetorically constructed, 53
Rhetorical nature of localization, 7
Rhetorical situation, 38, 150, 151, 205
Rhetoric of technology, 18, 92, 93
Risky elections, 96
Rough fingers, 81, 185, 196

S
Sahara Desert, 15, 41
St. Amant, Kirk, 167
Savage, Gerald, 21, 25, 54, 208
Savior of risky electoral system, 37
Scientific, 27, 55, 60, 106, 115, 116, 129, 136, 169, 170, 190, 194, 196
Security apparatus, 8
Seigel, Marika, 12, 22, 27, 68, 69, 71, 100, 108, 109, 111, 113, 140, 142, 148, 150–152, 158, 159
Self-determination, 30, 57, 75
Sense making, 137–139
Shoshana, Magnet, 115, 129
Silenced, 24, 26, 55–58, 60, 72, 124, 197
Site of struggle, 10
Situated action, 15, 131
Situatedness, 209, 214

Situatedness of technology, 16
Smith, Linda Tuhiwai, 55, 57, 58
Social, 1, 6, 9, 22, 27, 28, 38, 45, 47, 50, 53, 54, 58, 73, 89, 91, 92, 108, 110, 113, 115, 119, 124, 131, 132, 135, 148, 159, 173–175, 178, 179, 187, 189, 191, 195, 196, 202, 203, 205–207, 209, 214, 216
Social affordances, 9
Social justice, 21, 22, 26, 28, 30–32, 57–60, 70, 75, 182, 185, 186, 188, 191–195, 197, 208
Social learning process, 206
Sociocultural context, 7, 9, 16, 37, 38, 131
Socio-physiognomic, 193
Solomon Islands, 130, 186
Solve local problems, 7, 10, 192
Space, 3, 13, 37, 57, 58, 61, 106, 174, 186, 205, 206, 208, 212, 213, 216, 217
Spatial relations, 216
Spinuzzi, Clay, 16, 18, 19, 31, 100, 106, 146, 167, 168, 172, 192
Static representation, 54
Stepping back, 31, 137, 139–142
Stories of users, 79
Strategic use of language, 7
Subjugated, 56, 58, 60
Sub-Saharan Africa, 42
Subversion, 8, 31, 90, 146
Subversive localization, 10, 17–19, 29, 31, 75, 89, 146, 167, 168, 202
Subvert, 8, 18, 19, 31, 89, 146, 167, 168, 170, 172, 202, 207, 212
Successful elections, 42, 43, 53, 182, 204, 207
Successful use of technology, 113
Suite of posters, 28, 30, 69, 75, 93, 96, 103, 104, 108–110, 114
Sullivan, Patricia, 53

Sun, Huatong, 4, 6–10, 12, 13, 15–17, 23, 27, 54–56, 60, 131–133, 138, 143, 164, 167, 168, 188, 192, 203, 205, 208, 216
Super-hero narratives, 129, 188
SuperLock Technologies Limited (STL), 3, 93, 99, 104, 117, 130, 149, 164
Supreme Court of Ghana, 47, 100, 147, 204
Surveillance, 89, 114, 121
System constituting, 100, 108
System maintaining, 109, 159
System of ordering, 27

T
Tactical technical communication, 209
Tactics, 9, 22, 30, 31, 132, 143, 172, 207, 208
Techne, 113, 120, 169–172
Technical communication, 6, 7, 18, 20–25, 27, 54, 62, 91, 92, 103, 132, 137, 145, 146, 148–151, 165, 167, 182, 188, 189, 191, 192, 201, 202, 208, 209, 213, 214, 216
Technical communication practice, 23, 27, 208
Technical communicators, 11, 12, 20, 21, 23–28, 71, 92, 100, 115, 140, 149, 150, 187, 195, 202, 208, 209, 211–215
Technical discourses, 209
Technical documentation, 28, 29, 147, 212
Technical documents, 22, 26–28, 53, 68, 71, 90, 92, 93, 100, 103, 104, 108, 182, 209, 214
Technical genres, 30, 69, 100, 129, 202
Technical writing, 27, 28, 68, 109, 126, 145, 150, 215
Technological gap, 164
Technology accommodation, 27
Technology determinism, 16, 17, 132
Technology studies, 21, 23, 27, 202
Technology transfer, 11, 84, 201, 202, 209
Tension, 4, 132
Tension between localization, 7, 164
Textual analysis, 28
Thatcher, Barry, 54, 55, 67, 173, 175, 179, 201, 202, 205, 210, 211, 215
Timber, 32, 42
Togo, 41, 42, 45
Top-down, 5, 58, 92, 179, 188, 191
Transferability, 14
Translation, 8, 12, 16, 91, 92, 166, 192
Transparency, 50, 86, 89, 125
Triumph of users, 164
Trust in things technical, 18, 107

U
Unblackbox, 197, 209
Unenfranchised/disenfranchised cultural sites, 12, 24, 119, 165
United Nations (UN), 25, 43, 124
Universal Declaration of Human Rights, 38, 187
Use, 2–5, 7–11, 13, 15, 17–20, 22–24, 26, 27, 29–32, 38, 42, 44, 46, 48–50, 54, 56, 60, 64, 66, 67, 69, 71, 73, 79, 82–84, 88–90, 93, 96, 100, 103–106, 108–110, 112, 114–118, 120–123, 125, 126, 129–140, 143, 146–152, 158, 163, 164, 168–170, 174, 175, 178, 179, 185, 188–190, 192, 193, 195,

197, 202, 205, 207, 208, 211, 212, 216, 217
Use activities, 6, 7, 164, 165
Use Coca-Cola, 3
User actions, 131, 149, 189, 205
User advocacy, 24
User-centered design, 16, 21, 164
User-driven approach, 6
User experience, 4, 12, 17, 25, 130–132, 138, 203, 208
User heuristic experience, 130
User-heuristic localization, 10, 29, 75, 132, 139, 142
User in localization, 11
User knowledge, 16, 19, 23, 32, 135, 136, 143, 170, 192, 208
User localization, 15–17, 32, 132, 164, 203
User manual, 14, 19, 20, 22, 28, 30, 31, 49, 69, 73, 74, 90, 93, 99, 103, 104, 108–110, 145, 147–149, 151, 158, 159, 163, 164, 166, 168, 172–175, 178, 179, 182, 207, 212, 215, 216
Users, 3–13, 15–24, 27–31, 37, 55, 58–61, 68, 71, 75, 79, 81, 82, 84, 88, 89, 91, 92, 100, 103, 104, 108, 109, 111, 113, 123, 125, 131–140, 142–152, 158, 159, 163–170, 172, 179, 182, 185, 188, 189, 191–193, 197, 198, 202–209, 211, 212, 216
User satisfaction, 17
Users control design, 16
Users in localization, 10, 37, 57
User situations, 15, 16
Users save technology, 133
User strategies, 11, 15, 16, 25, 31, 75, 132, 136, 149, 202

Users' voice, 11
User tactics, 10
Use situations, 6, 7, 9, 16, 31, 140, 165, 167, 168

V
Vacca, J.R., 49, 50, 190
Value-neutral, 31, 56, 129
Verification, 2, 3, 19, 26, 28, 29, 38, 46, 49, 50, 69, 95, 96, 98, 102, 112, 113, 115, 118, 121, 122, 130, 134, 147, 177, 186, 198, 207
Verification officers, 20, 69, 70, 87, 147, 163, 181
Vernacular knowledge, 135, 136
Victims of technology, 12
Vote rigging, 2, 42–44, 47, 65
Voters register, 44, 180
Voting by minors, 2
Vulnerable population, 189, 195
Vulnerable technology, 195

W
Walton, Rebecca, 22, 27, 54, 67, 68, 187, 188, 191, 215
Wash hands, 3, 10, 73, 75, 85, 134, 137, 140
Weather conditions, 3, 15, 18, 41, 90, 130, 139, 141
West Africa, 37
Western cultures, 23, 57, 134
Wicked ambiguities, 4
Wicked problem, 21
Wicked question, 20